Machine Learning: Foundations, Methodologies, and Applications

Books published in this series focus on the theory and computational foundations, advanced methodologies and practical applications of machine learning, ideally combining mathematically rigorous treatments of a contemporary topics in machine learning with specific illustrations in relevant algorithm designs and demonstrations in real-world applications. The intended readership includes research students and researchers in computer science, computer engineering, electrical engineering, data science, and related areas seeking a convenient medium to track the progresses made in the foundations, methodologies, and applications of machine learning.

Topics considered include all areas of machine learning, including but not limited to:

- Decision tree
- Artificial neural networks
- Kernel learning
- Bayesian learning
- Ensemble methods
- Dimension reduction and metric learning
- Reinforcement learning
- Meta learning and learning to learn
- Imitation learning
- Computational learning theory
- Probabilistic graphical models
- Transfer learning
- Multi-view and multi-task learning
- Agents and Multi-Agent Systems
- Graph neural networks
- Generative adversarial networks
- Federated learning
- Large Language Models
- Multimodal Learning
- Transformer/Diffusion Models
- Generative Artificial Intelligence
- Bio-inspired Learning Models
- Embodied AI
- Explainable AI and Ethics

This series includes monographs, introductory and advanced textbooks, and state-of-the-art collections. Furthermore, it supports Open Access publication mode.

Huaping Liu • Xinzhu Liu • Kangyao Huang •
Di Guo

Embodied Multi-Agent Systems

Perception, Action, and Learning

Springer

Huaping Liu
Department of Computer Science and Technology
Tsinghua University
Beijing, China

Kangyao Huang
Department of Computer Science and Technology
Tsinghua University
Beijing, China

Xinzhu Liu
Artificial Intelligence Institute of CETC
Beijing, China

Di Guo
School of Artificial Intelligence
Beijing University of Posts and Telecommunications
Beijing, China

ISSN 2730-9908 ISSN 2730-9916 (electronic)
Machine Learning: Foundations, Methodologies, and Applications
ISBN 978-981-96-5870-1 ISBN 978-981-96-5871-8 (eBook)
https://doi.org/10.1007/978-981-96-5871-8

This Springer imprint is published by the registered company Springer Nature Singapore Pte Ltd.
The registered company address is: 152 Beach Road, #21-01/04 Gateway East, Singapore 189721, Singapore

All our knowledge begins with the senses, proceeds then to the understanding, and ends with reason.

— Immanuel Kant(1724–1804)

Preface

This book focuses on active perception and interactive learning for embodied multi-agent systems.

Active perception, particularly active vision, encourages adjusting the camera(s) viewpoint to explore the environment and gather more accurate information. This area of research is crucial as it bridges the gap between computer vision and robotics.

The basic idea of active perception (under the term of accommodation) was proposed in Tenenbaum's dissertation in 1970 and was formally formulated by Prof. Ruzena Bajcsy in 1988.[1] In another classical paper,[2] Prof. John Aloimonos pointed that ill-posed, nonlinear or unstable problems for a passive observer become well-posed, linear or stable for an active observer. Following the famous discussion that took place in 1994,[3] the research of active perception encountered some challenges due to the insufficient integration of sensor hardware with perception software. Consequently, computer vision researchers shifted their focus toward Internet data, neglecting active exploration of the physical world.

In spite of the great success of computer vision, Prof. Danica Kragic noted[4] that many existing solutions are not readily applicable to robotics. Robotic applications bring some important challenges that are not adequately or at all addressed in the computer vision community. In 2018, Prof. Ruzena Bajcsy, Prof. John Aloimonos, and Prof. John K. Tsotsos revisited the concept of active perception[5] and emphasized that active perception must be incorporated for a fully functional artificial agent.

---

[1] Bajcsy R. Active perception. Proceedings of the IEEE, 1988, 76(8): 966–1005.

[2] Aloimonos J, Weiss I, Bandyopadhyay A. Active vision. International journal of computer vision, 1988, 1: 333–356.

[3] CVGIP: Image Understanding, 1994, vol.60, no.1.

[4] Kragic D. From active perception to deep learning. Science Robotics, 2018, 3(23): eaav1778.

[5] Bajcsy R, Aloimonos Y, Tsotsos J K. Revisiting active perception. Autonomous Robots, 2018, 42: 177–196.

Recently, the development of embodied intelligence has sparked interest in the study of active perception. The embodied intelligence is widely observed in biological systems and emphasizes the collaborative effect of the brain, body, and environment. The behavior arises from continuous and dynamic interactions between the agent and its environment through perception and physical actions. Unlike disembodied intelligence, which focuses on logic, reasoning, and problem solving, embodied intelligence can develop in a complementary and synergistic way, paving the way for new insights into the nature of intelligence.

The study of embodied intelligence originated in psychology but has recently attracted significant attention in fields such as cognitive science, AI, and robotics. Specifically, the embodied cognition theory suggests that "the brain perceives the world through the body," highlighting that intelligence is influenced by the interplay of the brain, body, and environment. This viewpoint has paved the way for enhancing behavioral intelligence in robotics. Human-being can execute highly complex tasks in dynamic and uncertain environments due to the collaborative system of "brain-body-environment," which actively perceives and integrates various sensory inputs, such as sight, sound, and touch. This integration forms the basis for self-learning capabilities. Intelligent robots designed using the mechanisms of "brain-body-environment" collaboration exemplify AI-driven systems and embody the trends in next-generation robotics.

On the other hand, the remarkable reasoning, perception, and decision-making capabilities demonstrated by LLM in recent years have brought significant opportunities for the exploration of artificial general intelligence (AGI). This results in the development of increasingly larger models and a higher consumption of data. The ultimate goal is to achieve AGI through a unified brain model. However, when it comes to embodied agents, this strategy encounters considerable challenges due to the variety in morphology and function among these agents. It is neither feasible nor desirable to expect all embodied agents to conform to a single morphology. Instead, we should embrace the principles of biodiversity, promoting the existence, collaboration, and interaction of various forms. This recognition has motivated our research into embodied multi-agent systems. During this process, we have realized that active perception, along with the interactive learning capabilities that stem from it, plays a crucial role in fostering collaboration and synergy among multiple embodied agents.

This book discusses the recent progress of embodied multi-agent systems from the viewpoints of active perception and interactive learning. It is structured into four parts. Part I presents the motivation and background of embodied intelligence. Notably, Chap. 1 provides a comprehensive survey about embodied intelligence, and Chap. 2 focuses on embodied multi-agent systems. Part II focuses on theories and methods. In Chap. 3, the basic perception-action loop in embodied multi-agent systems is introduced, and Chap. 4 introduces the embodied cooperation in multi-agent systems. Chapter 5 presents the more advanced competitive learning in embodied multi-agent systems. Chapter 6 discusses the large language model for embodied multi-agent systems. Part III addresses applications, with Chap. 7 presenting a simulation platform for human-robot collaboration, and Chap. 8 demonstrating typical practical cases of embodied multi-agent systems. Finally, Part IV summarizes the findings and outlines potential future directions.

This book is suitable as a reference book for graduate students with a basic knowledge of machine learning, as well as for professional researchers interested in robotics and embodied intelligence.

Beijing, China
December 2024

Huaping Liu
Xinzhu Liu
Kangyao Huang
Di Guo

Acknowledgments

In the course of our investigation into active perception and interactive learning for embodied multi-agent, we received lots of support from many scholars, including Prof. Angelo Cangelosi, Prof. David Hsu, Prof. John Aloimonos, etc.

Throughout our research, we have benefited greatly from the assistance of industry experts. We would like to express our gratitude to Dr. Hongbo Li from Geek+ and Dr. Tianlei Zhang from TrunkTech for their invaluable support in our study of embodied multi-agent collaboration. Additionally, we would like to express our sincere gratitude to Chenxu Wang, Xinghang Li, Juan Wang, Peiyan Li, Pingcheng Jian, and Chuye Hong for their significant assistance in preparing the figures and proofreading the book.

The completion of this book is closely tied to the support from Springer Publishing. We are grateful to our commissioning editor, Dr. Lanlan Chang, for her exceptional assistance and support.

Much of the research work presented in this book is supported by the National Natural Science Fund for Distinguished Young Scholars under Grant 62025304.

Beijing, China

Huaping Liu
Xinzhu Liu
Kangyao Huang
Di Guo

Declarations

Competing Interests This study was funded by National Natural Science Fund for Distinguished Young Scholars under Grant 62025304.

The authors have no conflicts of interest to declare that are relevant to the content of this chapter.

Contents

Part I Background

Part IV Conclusions

Acronyms

A-V	Audio-Visual
AIGC	AI-generated Content
ALFRED	a new benchmark named Action Learning From Realistic Environments and Directives
API	Application Programming Interface
AR	Augmented Reality
BERT	Bidirectional Encoder Representations from Transformers
CNN	Convolutional Neural Network
CL	Contrastive Learning
COMA	Counterfactual Multi-Agent Policy Gradients
COT/CoT	Chain-of-Thought
CTDE	Centralized Training with Decentralized Execution
CV	Computer Vision
DDPG	Deep Deterministic Policy Gradient
DTD	Dynamic Task Decomposition
DTA	Dynamic Task Allocation
EAI	Embodied Artificial Intelligence
EI	Embodied Intelligence
EQA	Embodied Question Answering
FTA	Fixed Task Allocation
FTD	Fixed Task Decomposition
GAN	Generative Adversarial Network
GCN	Graph Convolutional Network
GPL	Graph-based Policy Learning
GPU	Graphics Processing Unit
HITL	Human-in-the-Loop
HER	Hindsight Experience Replay
HRC	Human-Robot Collaboration
HRL	Hierarchical Reinforcement Learning
IL	Imitation Learning
IR	Information Retrieval

IRoT	Interactive Reflection of Thoughts
LICA	Learning Implicit Credit Assignment
LLM	Large Language Model
LSTM	Long Short-Term Memory
MADPG	Multi-Agent Deep Deterministic Policy Gradient
MAAC	Multi-Actor-Attention-Critic
MAPPO	Multi-Agent Proximal Policy Optimization
MARL	Multi-Agent Reinforcement Learning
MASAC	Multi-Agent Soft Actor-Critic
MDP	Markov Decision Process
MLP	Multi-Layer Perceptron
MPC	Model Predicted Control
MR	Mixed Reality
NeRF	Neural Radiance Fields
NE	Nash Equilibrium
NLP	Natural Language Processing
PI	Physical Intelligence
PPO	Proximal Policy Optimization
PRC	Physical Reservoir Computing
PWSR	Path-weighted Success Rate
RC	Reservoir Computing
R-CNN	Region-based Convolutional Neural Network
RL	Reinforcement Learning
RNN	Recurrent Neural Network
ROBEL	Robotics Benchmarks for Learning
RSS	a famous academic conference named Robotics: Science and Systems
SAC	Soft Actor-Critic
SAVN	Self-Adaptive Visual Navigation
SLAM	Simultaneous Localization and Mapping
SPL	Success weight by Path Length
SR	Success Rate
TarMAC	Targeted Multi-Agent Communication
ToM/TOM	Theory-of-Mind
TWSR	Time-Weighted Success Rate
VDN	Value Decomposition Networks
VLN	Visual Language Navigation
VLM	Visual Language Manipulation
VR	Virtual Reality
VSN	Visual Semantic Navigation
VSR	Voxel-based Soft Robot

List of Figures

List of Tables

Part I
Background

This part of the book comprises two chapters. Chapter 1 presents a survey on embodied intelligence, and Chap. 2 addresses the motivation and technical challenges associated with embodied multi-agent collaboration.

Chapter 1
Embodied Intelligence

Abstract Embodied intelligence emphasizes that the intelligence is affected by the tight coupling of brain, body and environment. It is continuously and dynamically generated through the process of information perception and physical interaction with the environment. During the past years, the research scope of embodied intelligence has also been expanding and it has attracted great attentions from various communities. At the same time, a significant amount of research related to embodied intelligence has been introduced, particularly in the last few years. In this chapter, we present a comprehensive survey of embodied intelligence, viewing it as a synergy of morphology, action, perception and learning, along with a thorough summary and categorization of existing studies. Specifically, as the embodied intelligence is a synergy of all these components rather than themselves alone, we mainly focus on the connections across these four components (morphology, action, perception and learning) and identify areas where future research can benefit from their intrinsic connections.

1.1 Introduction

Embodied Intelligence (EI) is the computational approach to the design and understanding of intelligent behavior in embodied and situated agents through the consideration of the strict coupling between the agent and its environment, mediated by the constraints of the agent's own body, perceptual and motor system, and brain [23]. Embodied intelligence emphasizes that the intelligence is affected by the tight coupling of brain, body and environment. It is continuously and dynamically generated through the process of information perception and physical interaction with the environment. Embodied intelligence is the basis for living things to survive in complex environment, and it widely exists in plants, animals and other biological organisms. Held and Hein validated the relationship between vision, development and movement through the famous kittens experiment [69]. During the experiment, two neonatal kittens were kept in a room together. One was free to move actively, and the other was forced to do the equivalent movements passively with a specifically designed apparatus. Although the two kittens received the same visual

H. Liu et al., *Embodied Multi-Agent Systems*, Machine Learning: Foundations, Methodologies, and Applications, https://doi.org/10.1007/978-981-96-5871-8_1

stimulation, after some time, it was observed that the active kitten had developed a normal sensory-motor system, while the passive one had shown a great sensory-motor impairment. This experiment strongly verifies the importance of embodied learning.

In this chapter, we discuss the Embodied Artificial Intelligence (EAI), which is actually not a new concept. As early as the 1950s, Alan Turing points out that one way to build an intelligent machine is to provide it with the ability to perceive the world and let it develop by itself [184]. In classic artificial intelligence books and textbooks, the definition of an agent is *perceiving its environment through sensors and acting upon that environment through actuators* [159]. Therefore, the interaction between the body and its environment has consistently been a crucial method for exploring, simulating, and enhancing intelligence. Currently, the artificial intelligence technology, bolstered by the big data, deep learning, and graphics processing unit (GPU), has achieved unprecedented success in fields such as visual perception, language processing, and speech processing. Nonetheless, the area of robotics that emphasizes action execution has not yet reached the same level of success. Furthermore, even in the aforementioned areas where great success has been achieved, this success is primarily based on the assumption of a restricted environment. When it comes to open environments, current artificial intelligence techniques are still faced with huge challenges. The core reason for this problem is that we have paid more attention to the perception and learning capabilities of the agent, while not fully investigating the action and behavior capabilities of the agent. EAI is proposed to emphasize the role of the body, particularly its impact on intelligent behavior. On the contrary, the intelligent technology that ignores the body's influence is referred to as "disembodied intelligence". It is noted that actions are not completely ignored in the study of disembodied intelligence; however, there is generally a greater focus on simulating brain functions to make decisions and perform actions, such as in chess and computer games. Also, while the body is not completely overlooked in disembodied intelligence research, it is typically only utilized to passively carry out commands.

During the past several years, there has been a growing number of tutorial and survey papers addressing embodied intelligence from various perspectives. Here we make a brief analysis and clarify the motivation of this chapter. Reference [153] originally discloses the fact that higher-level cognition and intelligent behavior need to be understood from an bottom up manner, and an embodied intelligence architecture is established. Since then, the embodied intelligence has attracted more and more attention. Especially in recent years, embodied intelligence becomes quite popular in fields of robotics, computer vision and machine learning. The current survey papers can be mainly classified into two categories. The first category closely aligns with the research trajectory of [152, 153]. Reference [78] provides a review of embodied intelligence from the perspective of timescales, exploring how interactions function within complex autonomous and adaptive systems. Reference [173] discusses the physical intelligence (PI) that is inherent in the body, while embodied intelligence focuses on the tight coupling across an agent's body, brain, and the environment. It should be noted that physical intelligence also facilitates action generation, perception, and learning through the body, but it does not address its relationship with the brain and environment. The second category pays more attention on the recent progress highly related to machine learning, computer vision, and fundamental models, in which some more specific research topics such as visual language navigation, fundamental models, and simulators are summarized [46, 52, 121]. Reference [115] provides a more comprehensive review on EAI, while it neglects the connection between morphology and action. Different from the above two categories, Ref. [157] systematically discusses the inductive biases for embodied intelligence, analyzing the roles of morphology, representation and learning. It also describes the energy exchange between agent and the environment. However, the primary aim of Ref. [157] is to emphasize the challenges and opportunities within embodied intelligence and to suggest research directions that could significantly enhance the current state-of-the-art in robot learning. In this chapter, the embodied intelligence is regarded as a synergy of morphology, action, perception and learning. And a comprehensive survey and categorization of existing studies are presented. Our main focus is on the interconnections among these four components (morphology, action, perception, and learning), and we propose a unified embodied intelligence framework that considers the tight coupling of body, brain, and environment.

1.2 A Long Past and a Short History

The embodied intelligence has a long past, while a short history. A series of works by *Aristotle*, *Charles Darwin*, *Merleau Ponty*, *Claude Bernard*, *Walter Cannon*, *Martin Heidegger*, *Norbert Wiener*, *Jean Piaget*, *James Gibson* and others have laid a solid foundation for the study of embodied intelligence in philosophy, physiology, psychology, and cognitive science. The most representative viewpoint at present is that the brain plays an important role in the embodied intelligence, but it is not the origin of cognition. Since there is a dynamic loop between the perception and action, the perception and cognition establish an interdependent relationship through the body environment interaction. Therefore, the perception-action loop is actually the core of the cognition. This framework blurs the distinction between the perception and cognition. This is consistent with what is mentioned in [8] as *once we begin exploring the actual mechanisms that animals use to negotiate their worlds, it becomes very hard to decide where perception ends and cognition starts.*

On the other hand, the modern artificial intelligence originated from the Dartmouth conference in the 1950s, and for a period of time thereafter, the study of artificial intelligence is largely restricted to the paradigm of symbol processing (also known as symbolism). The limitation of this method is quickly observed in practical applications, leading to the development of connectionism, which includes multilayer perceptrons, feedforward neural networks, recurrent neural networks, and deep neural network methods that are popular in both the academia and industry today. This method of using neural networks to simulate cognitive science has indeed made great progress in aspects of adaptation, generalization, and learning, but it has not solved the challenging problem in the interaction between agents and the real physical world, and it also exhibits great limitations in the interpretability and other aspects.

In fact, the fundamental problem of symbolism and connectionism is that they ignore the interaction with the physical world. *Moravec's Paradox* which was originally proposed in 1980s points out that it is relatively easy for a computer to play chess like an adult, but it is quite difficult or even impossible for a computer to perceive and act like a one-year-old child. Interestingly, when we look at the flourishing achievements of artificial intelligence today, it is indeed still difficult for us to find a fundamental breakthrough to this problem. *Rodney Brooks* once puts forward that intelligence requires a body, emphasizing that intelligence is embodied and closely related to the environment. This viewpoint promotes the birth of behaviorism that is represented by the embodied artificial intelligence. It is noted that the "embodied artificial intelligence" mentioned here is a simulation of embodied intelligence that is achieved by artificial intelligent approaches. It is different from the inherent embodied intelligence of living creatures. We focus on the embodied artificial intelligence in this book. Unless otherwise specified, the "embodied intelligence" and "disembodied intelligence" mentioned in the subsequent content of this book refer to "embodied artificial intelligence" and "disembodied artificial intelligence" respectively.

Embodied intelligence and disembodied intelligence are not mutually exclusive. Both embodied intelligence and disembodied intelligence can be found in the symbolism, connectionism, and behaviorism, while each has its own focuses. They also share the same goals, including *understanding biological systems*, *imitating intelligent behavior* and *designing artificial agents*. Driven by the big data, deep learning and graphic processing units, the disembodied intelligence has achieved great success in fields represented by the Internet information processing, while the embodied intelligence highly relevant to mechanisms and materials has become the core foundation of intelligent robots. In the process of exploring intelligence, embodied intelligence and disembodied intelligence should complement to each other and develop cooperatively to promote the understanding, simulation, and expansion of the intelligence jointly. The close combination of the embodied intelligence and disembodied intelligence is a necessary way to realize the general artificial intelligence.

1.3 Architecture of Embodied Intelligence

In this section, we formulate a unified embodied intelligence framework which is composed of morphology, action, perception and learning modules (Fig. 1.1). As the embodied intelligence is a synergy of all these modules, we pay more attention to the connections between these modules (links in Fig. 1.1) rather than the modules themselves. To this end, eight subsections which correspond to the links in Fig. 1.1

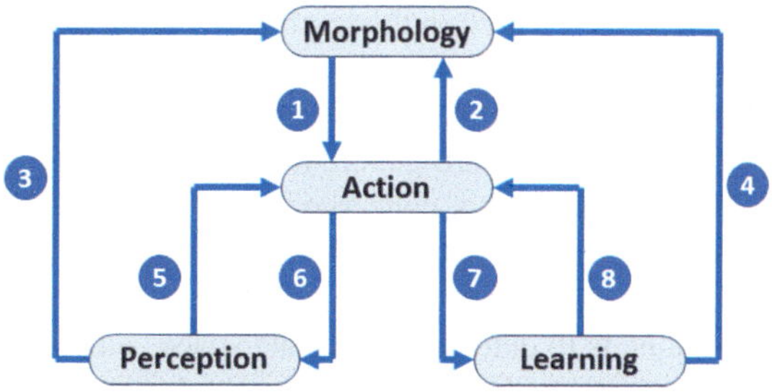

Fig. 1.1 The architecture of the embodied intelligence

are introduced in this section. Since Link ⑧ (from Learning to Action) in Fig. 1.1 is in fact the basis of many existing work, we introduce it first. And then the other links will be introduced in accordance with the sequential order.

1.3.1 Action Generation Based on Learning ($\mathcal{L} \rightarrow \mathcal{A}$)

This part corresponds to the link ⑧ from *Learning* to *Action* in Fig. 1.1, which focuses on using machine learning methods to generate actions. Actually, the action generation problem has been widely investigated in both the classical and modern control theories for several hundreds years. In the classical control area, the most widely employed PID controller works by adjusting the three parameters of proportional term, integral term and differential term. It is quite convenient and effective. However, the parameter adjustment requires rich engineering experience, which prevents such method to be applied for more complicated scenarios. And then, in the 1950s, modern control theory based on state space model was established. This kind of methods highly depend on the white-box model to get the controller parameters and complete stability analysis can be achieved. However, for complicated robotic and process control systems, it is difficult to obtain the accurate models and there always exist great gaps between the mathematical calculation and practical application. All these limitations motivate people to think about introducing the artificial intelligence technology to develop new intelligent control methods, and the reinforcement learning methods become more and more popular.

The basic idea of reinforcement learning should be attributed to the Markov decision process (MDP) established by Richard Bellman in the 1950s [9], with which an agent selects a best action based on its current state. On this basis, the Q-learning algorithm proposed by Chris Watkins [196] in 1989 lays the foundation for reinforcement learning. However, reinforcement learning also exhibits limitations such as convergence difficulty, poor generalization capability, etc. On the other hand, with the development of deep learning, the Deep Q-network [126] has seamlessly combined the feature extraction ability of convolution neural network and the action learning ability of reinforcement learning together, and great success has been achieved in the human-computer chess game, which once again inspires a new wave of research in the reinforcement learning. Recent studies have shown that reinforcement learning is also widely used in biological intelligence [133].

At present, reinforcement learning methods have become important approaches for the embodied intelligence. It is noted that the design of reinforcement learning algorithms is what is concerned by the field of machine learning [187]. We can directly use off-the-shelf reinforcement learning algorithms such as PPO [167] and SAC [65] that are most widely used to generate actions for the agent. What needs to be done is to determine the state space, action space and reward function, which are closely related to specific tasks. In this sense, the $\mathcal{L} \rightarrow \mathcal{A}$ is the basis for other links in Fig. 1.1.

1.3.2 Action Generation Based on Morphology ($\mathcal{M} \rightarrow \mathcal{A}$)

This part corresponds to the link ① from *Morphology* to *Action* in Fig. 1.1, which emphasizes the morphological computation. By means of the morphological computation, part of the computation that needs to be done by the brain can be offloaded to the body when interacting with the environment. The body is used to improve computation efficiency and further control the behavior of the body. This phenomenon is very common in our daily life. For example, a person can easily grasp an object without carefully computing the motions and contact forces. Recently, with the development of mechanisms, soft robots and other fields, morphological computation is embracing new opportunities [71, 137], and has already shown great advantages in Sim2Real transfer and low-power computation [23].

1.3.2.1 Typical Examples

One typical example of using morphological computation to achieve the control of more complex behavior is the passive walking robot (Fig. 1.2), which was developed in the 1990s [36, 124]. By designing a special morphological structure, the robot can achieve a steady gait on a shallow slope without active control input. This work

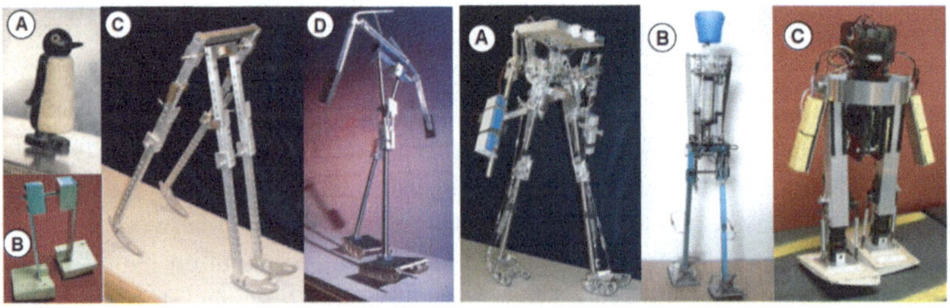

Fig. 1.2 Passive walking robots [36]. (Reproduced with permission from Ref. [36])

abandons the strict requirements on robot joint control in the dynamics modeling and control, and enables the robot to control its behavior completely relying on its own morphology.

Recently, there emerge more novel morphological structures. Reference [51] explores the use of morphological computation to achieve fast-running for quadruped robots. Reference [102] utilizes dynamic morphological computation to generate periodical gaits for snake-like robots. Reference [10] illustrates the morphological computation in haptic sensation from the aspects of display, sensing and interaction in both natural and artificial systems. Inspired by desert locusts, Ref. [201] studies the control method of animal legs adapting to different grounds, and carries out robotic dynamic adhesive experiments on surfaces including glass, sandstone, wood and mesh.

On the other hand, since increasing morphological computation for a robotic system potentially decreases the overall energy demand and the controller complexity, an energy viewpoint is established to characterize robots as the exchange of energy and information between a machine and its environment toward some specified set of goals [94]. This in turn prompts us to develop morphological computation for energy harvesting and other benefits. For example, Ref. [58] evaluates morphological computation in muscle of hopping movements, and Ref. [91] exploits the interconnection between a robot's adaptive morphology, control, and consequent behaviors to verify that the tight coupling between the brain and the body can increase energy efficiency via harvesting [170], which is worthy for further investigation.

Generally, works in this area are mainly closely relevant to mechanism design, and mostly the human interference is involved. Reference [53] presents a formal analysis method for morphological computation from the perspective of programmable dynamic systems. It points out that morphological computation is not only suitable for robots, but also can be widely used in scientific fields such as chemical systems, statistical physics, etc. However, a systematic guidance for morphological computation from the perspective of artificial intelligence technology is still lacking. Due to the great potential of morphological computation, many scholars have been actively exploring this area. Although a complete theoretical model for the morphological computation is not established yet, there emerge two relatively mature categories of methods: reservoir computing method and information theory method.

1.3.2.2 Physical Reservoir Computing for Morphological Computation

The most commonly used method in the morphological computation is the Reservoir Computing (RC) [80], which belongs to the dynamic system analysis method. In the RC model, the physical morphology is regarded as a Physical Reservoir Computing (PRC) device [130]. The computing mechanism of the RC model is a neural network structure in which the parameters of middle layers are randomly fixed and only parameters of the output layer need to be trained. Therefore, it is very

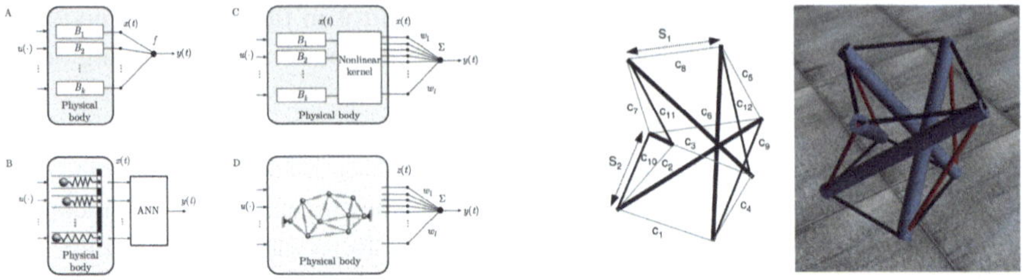

(a) (b)

Fig. 1.3 (**a**) Typical reservoir computing model [67]; (**b**) a 24-DoF tensegrity robot [148]. (Reproduced with permission from Ref. [148])

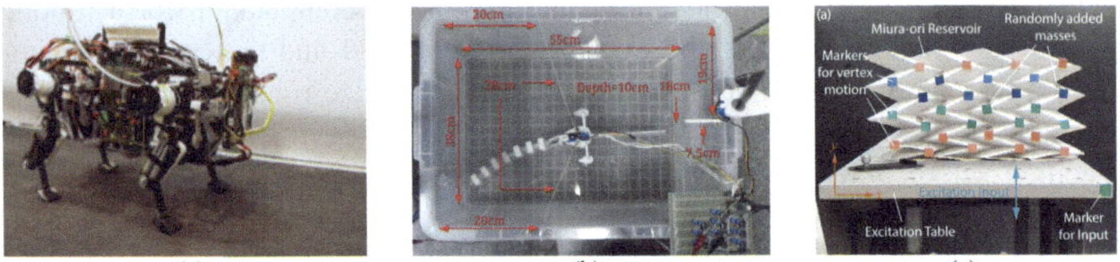

(a) (b) (c)

Fig. 1.4 (**a**) Quadruped robot [41]. (Reproduced with permission from Ref. [41]) (**b**) fish robot [73]; (**c**) a paper-based structure [12]

convenient to train the network as only a linear optimization problem needs to be solved. Since RC has a strong ability to describe dynamic systems, it has become a powerful tool for analyzing morphological computation.

On the other hand, since the internal parameters of the reservoir do not need to be explicitly learned, it is possible to leverage the dynamics of the physical body to implement the reservoir computation. Reference [67] constructs a reservoir computing model using mass-string systems, and its dynamic characteristics are analyzed in detail, providing a general reservoir computing model for morphological computation (Fig. 1.3a). Therefore, the follow-up works are mainly to build different physical reservoir computing models to realize the morphological computation. In this sense, the research on the reservoir computing model can effectively guide the design of the morphological computation device. A typical example is that if the mass mentioned above changes to a rigid rod, a tensegrity structure can be formed (Fig. 1.3b). This is a stable structure in which rigid elements are connected by tensile elements. Reference [148] uses four actuators to control a tensegrity robot with 24 degrees of freedom, demonstrating the computing ability of the tensegrity. Recently, Ref. [149] presents a unified framework for mechanical modeling and control of tensegrity robots.

In a quadrupedal robot gait control task (Fig. 1.4a), it is found that it is difficult to achieve complex morphological control by directly using the simple linear layer in the traditional reservoir computing model, and thus a nonlinear layer is

Fig. 1.5 Object detection in water environment [89]. (Reproduced with permission from Ref. [89])

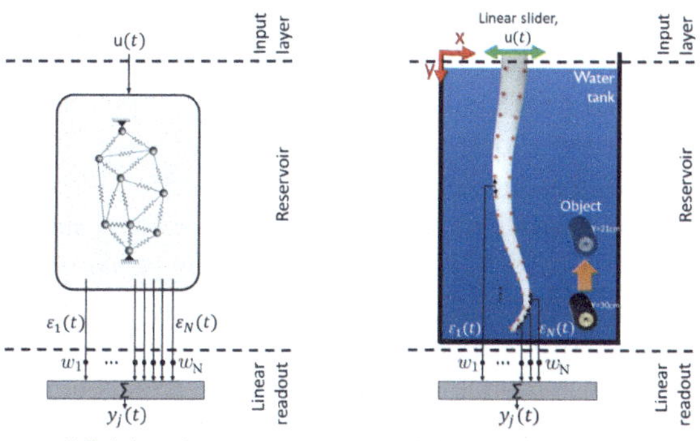

(a) Physical reservoir (b) Tentacle as a physical reservoir

introduced to further improve the performance of the reservoir computing model [41]. Reference [73] verifies that a fish robot system also meets the requirements of the echo state property in the reservoir computing model, and the morphological computation capability can be significantly improved by taking advantage of the swimming motions of the fish robot (Fig. 1.4b). Reference [12] points out that origami structures can also have sufficient dynamic performance to have sufficient morphological computation power to emulate high-order nonlinear systems, and a novel robotic crawling strategy is achieved (Fig. 1.4c). Besides, Ref. [129] simulates the morphological computation capabilities of human skin using the reservoir computing, and Ref. [182] uses the natural dynamics of the robot to recognize the wind direction. Recently, the reservoir computing, especially the physical reservoir computing develops rapidly, in which the reservoir computing is realized by physical devices, which is highly related to the embodied morphological computation. Reference [130] give detailed reviews of the recent progress in physical reservoir computing. In addition to physical reservoir computing, the current methods of using various physical morphology to realize neural network learning have also attracted great attention from researchers from different fields.

Furthermore, soft robots have attracted more and more attention in recent years and great progress has been achieved. Compared with rigid robots, soft robots have better characteristics in deformability, flexibility, safety, etc, and they are more suitable for driving on complex roads and manipulating unknown objects [88]. It is noted that the control of soft robot is very challenging due to the complex dynamics of its body, while from the perspective of the embodied morphological computation, such complex dynamics are instead valuable computing resources [108], and thus the soft robot becomes an ideal tool for the reservoir computing. Inspired by the octopus, Ref. [131] uses the octopus arm as a reservoir computing device, which is able to emulate complex nonlinear behaviors by reading a linear and static output from its physical body. And without an external controller, a closed-loop control can be achieved by converging to limit cycles. Based on this work, Ref. [132] demonstrates the approximation and control capabilities of this class of soft robots for nonlinear continuous functions. Reference [89] further uses this mechanism to realize the object localization in the water (Fig. 1.5). In addition, Ref. [47] expands the scope of embodied morphological computation with soft hand, and a reservoir model for pneumatically driven soft hand is developed. Reference [208] points out that the wrinkles produced by soft materials after soaking in water for a long time can help realize some computation tasks and it is applied to the active tactile perception. Reference [136] uses a soft whisker sensor for morphological computation to realize the active distance estimation. Reference [45] studies the method of tuning the damping property of soft silicone gripper, realizing the dynamic morphological computation. It can be seen that soft robots, as a very promising robot morphology, have become an important development direction of embodied morphological computation. It is believed that the development of soft robots will definitely drive a new wave of embodied morphological computation.

1.3.2.3 Information Theory Method for Morphological Computation

Although methods based on dynamical systems, especially the reservoir computing models have achieved great success, such methods are mainly aimed at qualitatively simulating embodied morphological computation and it lacks the quantitative evaluation ability, which limits further analyzing the mechanism of morphological computation in depth, while the information theory methods have obvious advantages in this respect. In fact, the most important thing to evaluate the performance of a morphological computation system is to analyze how much of the computation that should be undertaken by the "brain" is offloaded to "body". Inspired by the above idea, Ref. [57] points out that the quantitative analysis of morphological computation requires the establishment of a causal model of the cognitive system including the brain, actuators, sensors, and environment.

In addition, there are also some works to evaluate morphological intelligence from the perspective of analyzing control complexity. For example, Ref. [158] utilizes probabilistic optimal control methods to analyze how much computational work can be undertaken by the morphology of the robot by optimizing the controller complexity. Reference [37] compares the entropy of controllers corresponding to different morphology to analyze how the morphology affects behaviors. Reference [150] investigates to use entropy to characterize the embodiment of robots on the microscale. Cheap control also provides an alternative for morphological intelligence in which the system must fully exploit the role of embodied intelligence in the pursuit of cheap control. Reference [155] further introduces this idea into reinforcement learning, and a new optimization objective is established which encourages the morphological computation. Compared with the framework in [57], these works analyze the role of the morphology by indirectly analyzing the complexity of the controller. More recently, Ref. [48] indicates that simplicity is typically quantified on an action-by-action basis. And simple actions can be obtained by constraining the mutual information between states and actions, which satisfies the complexity of policies. It is noted that most of these quantitative analysis works are theoretical research independent of the actual physical morphology. There are still great challenges in how to combine specific physical morphology to achieve quantitative analysis of the morphological computation.

1.3.2.4 Summary

In general, this area of research work mainly investigates how to offload the computation tasks that the "brain" is responsible for to the "body", and how to accurately evaluate the morphological computation undertaken by the "body". It is a novel interdisciplinary integrating the mechanism, material and control technology. In addition to artificially designing morphological structures, great progress has been made in the theory of morphological computation, including reservoir computing models and information theory analysis methods. The "Curse of Dimension" problem that is common in the field of high-dimensional robot control may be effectively overcome by the advantages brought by the morphology, which we also call as "Blessing of Morphology".

1.3.3 Morphological Control Based on Action ($\mathcal{A} \rightarrow \mathcal{M}$)

This part corresponds to the link ② from *Action* to *Morphology* in Fig. 1.1. The morphological structure is resorted to simplify the control for certain actions. For the embodied agent, no matter how complex its morphology is, it has certain "structural" characteristics. Such structural information can help to reduce the optimization space for action generation, guaranteeing the feasibility and explainability of the solution.

When designing a controller for agent with complex morphology, traditional reinforcement learning does not take advantage of the morphological characteristics of the agent, but just simply concatenates observations from different parts of the agent, and directly outputs all control variables. This results in a large control variable searching space and the learned control variables are difficult to transfer between different morphologies. The main task of action-based morphological control is how to effectively integrate the morphological information into the learning process of the controller for the action. The most straightforward idea is to directly encode the agent morphological information in order to solve the morphological control problem. Reference [31] proposes an explicit encoding and an implicit encoding methods respectively. For the explicit encoding, the relative poses of different joints are concatenated into a vector, which can facilitate modeling the kinematic structure of the robot. This encoding method utilizes some prior information, while it is limited to the simple concatenation operation, and is only suitable for some manipulators, but not more complex morphologies. For the implicit encoding, the encoding of the morphology is optimized during the process of optimizing the learning strategy iteratively, which means that the algorithm not only searches for the optimal mapping from the state to action, but also discovers the optimal encoding of the morphology. Although it shows good performance in the transferring ability between the same type of manipulators with different degrees of freedom, it brings new difficulties to the optimization, and the robot's prior morphological information is not fully leveraged.

At present, more advanced technologies such as the graph neural network and Transformers [185] are widely used to efficiently represent the morphological information and directly used to improve the efficiency of the reinforcement learning. More importantly, controllers that directly considering morphological information shows good performance for cross platform transferring, and has significant advantages when controlling new embodied agents.

1.3.3.1 Graph Neural Network for Morphology Control

In the field of reinforcement learning, there are many works that introduce graph structure to improve learning efficiency. But most of them are limited to using graph structure to describe the environment where the agent lives, rather than the morphological structure of the agent [161, 169]. Reference [192] establishes a basic morphological graph model method where the nodes represent different joints and edges represent the dependencies between them. With the graph model, we can directly use the message passing mechanism of the graph neural network to handle the relationship between various joints. During each learning iteration, each node can distribute its state information to adjacent nodes, and each node synthesizes the state information sent by other nodes to update its own state. In this way, the morphological information is effectively incorporated in the learning process, which facilitates the controller to learn more quickly. On this basis, Ref. [16] introduces parameter freezing techniques to train graph neural networks to solve high-dimensional continuous control problems. Although the method proposed by [192] effectively utilizes the morphological information, different strategies have to be designed for different nodes in the graph structure which corresponds to different parts of the agent. Reference [77] proposes a unified control policy for different morphological agents using graph structure. Its core idea is to decompose the morphology of the agent into different independent modules, and the essence is still to rely on the message from adjacent nodes to affect the policy. In this way, the control policy can be independently designed for different modules, and a unified controller can be achieved for different morphologies. It also provides insights for the pre-trained model for controller design. Recently, Ref. [199] uses the graph neural network to describe the structural information between modules of the agent, pointing out that modules of similar structure can share control policies. By integrating this structure information into the model-based reinforcement learning, the search space is significantly reduced and the result is verified on physical experiments (Fig. 1.6).

Overall, the original intention for using graph neural networks for morphology control is that the morphology information can generate an inductive bias that is beneficial for learning the controller. However, currently, the degrees of freedom

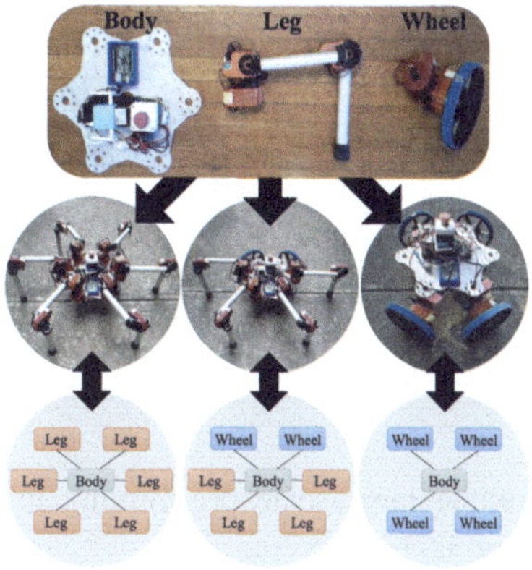

Fig. 1.6 Modular structure design for GNN description [199]. (Reproduced with permission from Ref. [199])

and types of robots that can be handled are still relatively limited. Further research is needed to accommodate efficient transfer between significantly different morphologies. Generally, for agents with simple morphological structures, there is little difference between different morphological description methods. For agents with complex morphologies, the graph structure methods show superior performance compared to those with simple morphologies. In practical usage, it is necessary to use the graph model to accurately capture the morphological characteristics of the agent.

1.3.3.2 Transformer for Morphology Control

Although the graph model plays a positive role in the morphology control, the morphology of the robot is generally a sparse graph structure, and the key information is likely to disappear after multiple rounds of message passing, which is the so-called "over-smoothing" problem. In recent years, Transformer has received great attention as a model based on self-attention mechanism [185]. If the attention is designed as an edge-to-vertex aggregation unit, the Transformer can be regarded as a graph neural network on a fully connected graph. Reference [100] directly uses Transformer to achieve message passing between different components, while the encoder and decoder can act on different components separately. Although this approach demonstrates the potential of the Transformer, it ignores the real physical morphology of the robot. Reference [72] further reveals the role of node's positional information on the self-attention mechanism, and the morphological information is embedded into the Transformer model for a joint policy learning of heterogeneous morphologies. It overcomes the over-smoothing problem caused by the sparse structure in conventional graph neural networks. Further, aiming at the large-scale robot morphology control problem, Ref. [61] considers the robot morphology as the Transformer input, and learns a universal policy by building a "Metamorph" to control a large number of different robots at the same time, laying the foundation

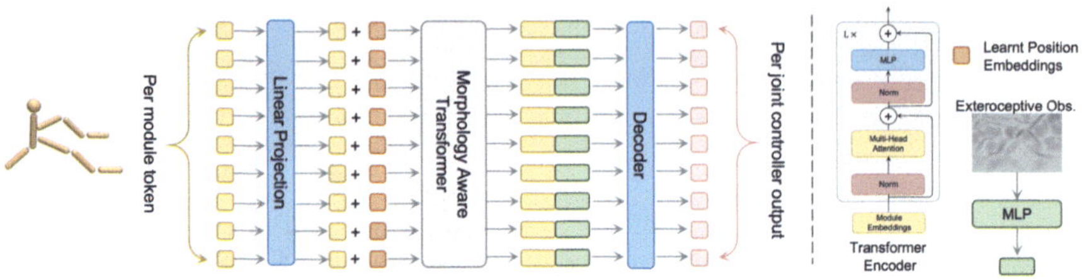

Fig. 1.7 Metamorph structure [61]

for the large-scale pre-training model for the embodied morphological learning (Fig. 1.7).

1.3.3.3 Morphology Transfer

For the morphology control transfer between different morphologies, as graph neural network and similar methods describe the characteristics of the morphology directly, good transferring performance for agents with similar morphologies are observed. However, when large morphological differences exhibit, the transfer performance deteriorates because that the corresponding state space and action space are quite different among different morphologies. To this end, Ref. [68] proposes a method that hierarchically decouples policies, where only high-level policies are transferred, while low-level policies are still learned independently. Due to the coupling of high-level policies and low-level policies, if the low-level policies of agents with different morphologies are significantly different, the transfer of high-level policies is also difficult to succeed. Therefore, the mutual information is introduced to minimize the difference between the morphologies and low-level actions, so as to achieve the alignment of the low-level policies of different agents. Overall, most current works are still in the simulation stage, and how to implement them in real robots, especially bionic robots and humanoid robots, is still a very challenging problem.

1.3.3.4 Summary

Due to factors such as high dimensionality, nonlinearity, and strong coupling, it is very challenging to use reinforcement learning methods to achieve morphology control for complex agents. However, the morphological information provides very important and useful prior information that can significantly constrain the searching space. Therefore, it is important to introduce morphological information into the learning algorithm in an appropriate way so as to improve the learning efficiency of the controller, and also ensure the performance when transferring the control policy to other unknown morphologies, reflecting the Blessing of Morphology. At present, the graph neural networks and Transformer are representative methods used to describe morphology structures. These works are still mainly conducted in the simulation environment, and the learned policies will encounter many difficulties when transferring to physical systems. On the other hand, although it is very attractive to design a unified morphological controller for different robots, it is still faced with a lot of difficulties. Inspired by the pre-training models in the fields such as natural language and computer vision in recent years, a unified pre-training large morphology control model for different morphologies is also an important research direction in the future.

1.3.4 Perception-Driven Morphology Transformation ($\mathcal{P} \rightarrow \mathcal{M}$)

This part corresponds to the link ③ from *Perception* to *Morphology* in Fig. 1.1, in which the agent morphology transforms online according to the environmental perception information. The morphology transformation is quite common in our lives. For example, when people skate fast on the ice, they will instinctively lower their body to maintain the balance, and the birds will quickly flap their wings and fly away after being thrilled. This morphology transformation is also often seen in the field of robotics, especially for amphibious robots and transformable robots, which can change their body structure or parameters adapting to the environment and tasks driven by the perception.

As the online morphology transformation is coupled with the perception information, a high real-time performance is required. Recently, Ref. [171] reports a multi-modal mobility robot which uses the method of changing appendages to enhance its movement ability (Fig. 1.8 LEFT). After transforming its wheels, legs and propellers, it can be switched to an unmanned ground vehicle, mobile inverted pendulum, unmanned aerial system, etc. It relies on the morphology transformation to traverse among different terrains. In addition, the quadruped robot proposed in [139] can dynamically adjust the length of its legs when walking on different grounds to adapt to different terrains. Another typical example is the amphibious fly-drive vehicle which is reported in [177]. Reference [93] presents a chameleon soft robot which can mimic the chameleon to adjust its color according

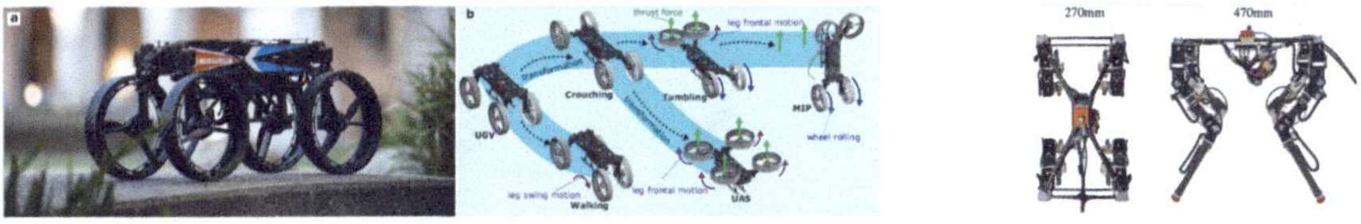

(a) (b)

Fig. 1.8 (**a**) A multi-modal mobility robot [171]; (**b**) Quadruped robot evolving system [139]. (Reproduced with permission from Ref. [139])

to the environment. Reference [202] uses thermal, chemical or electrical stimuli to continuously drive the movement of liquid metal droplets and makes the wheeled robot change its center of gravity for generating a rolling torque.

1.3.4.1 Summary

Perception-driven morphology transformation is highly related to the hardware. Though there are many related works, standardized design principles are still lacking. In addition, since the morphology transformation highly depends on areas of materials and mechanisms, its practical implementation are still faced with great challenges.

1.3.5 Learning-Driven Morphology Optimization ($\mathcal{L} \rightarrow \mathcal{M}$)

This part corresponds to the link ④ from *Learning* to *Morphology* in Fig. 1.1, which focuses on using learning technologies to realize morphology optimization. In biology, the morphology optimization phenomenon is common. The morphologies of creatures are optimized to adapt to the environment during the evolution. For example, limbs are evolved from fins, and after the morphological evolution tends to be stable, a better upright walking behavior is learned. In addition, the morphological parameters could be further optimized through acquired exercising. Inspired by this, we can search for a morphology that is more suitable for control policy learning through co-optimization. The joint optimization of the morphology and control policy for the embodied agent can help to accomplish the task more efficiently. As is shown in Fig. 1.9, the morphology of the agent's legs are optimized while learning to walk over obstacles.

Reference [134] points out that humans are better at designing physical systems than the intelligent control system. The idea of brain-body co-evolution based on

Fig. 1.9 Embodied agents with different morphologies walk over obstacles

learning has received great attention in the area of embodied intelligence since the early days [151], and it is sometimes called developmental robots, artificial life, etc. Reference [172] firstly uses the evolutionary learning framework to realize the co-optimization of morphology and controller in simulations. A three-dimensional rigid robot is represented as directed graph genotypes, and the evolutionary algorithm on the graph is used to optimize the morphology of the robot. These works has received widespread attention as they are able to help design robotic morphologies better adapting to the environment and tasks [4, 19, 75, 106, 118, 123, 174]. Reference [5] points out that the co-evolution of morphology and control is similar to the brain-body synergy, but ignores the influence of the environment. It verifies the effect of environmental complexity on morphological complexity. Although some progress has been made in the co-optimization of morphology and control at the beginning of this century, the optimization process is limited to the software and hardware simulation conditions, and no major breakthroughs have been made. In the past ten years, with the rapid development of manufacturing technology, graphics simulation and rendering technology, and computing technology, great achievements have been made in methods of co-optimization of morphology and control.

Currently, the work in this area mainly includes the optimization of morphological parameters and structure, and the methods used mainly include reinforcement learning [62, 64] and genetic optimization [70]. In recent years, there have also been works that simultaneously optimize the structure and parameters using only a reinforcement learning. The problems that need to be solved in this area are not only the efficient representation of the morphology, but also the efficient optimization of the algorithm.

1.3.5.1 Evolutionary Reinforcement Learning Methods

The early co-optimization problems of morphology and control are mainly based on the evolutionary search method, in which the evolutionary search space is large (including the search of morphological parameters and controller parameters). In recent years, work in this area has mainly focused on the study of different encoding methods for specific task requirements [34, 96, 142]. Although the morphology and control should be jointly optimized, they are actually optimized in different scales. Taking the creature as an example, the change of morphology is more like an evolutionary process, while the design of the controller is more like the acquired learning process given a specific morphology. Therefore, a very natural idea is to use the evolutionary optimization method to optimize the morphological structure and parameters, and use the reinforcement learning method to realize the optimization of the controller parameters. Based on the graph neural networks controller proposed in [192], Ref. [193] describes the robotic morphology design as a graph search problem. It introduces the concept of species, and designs mutation operations to add and delete nodes, enabling the evolutionary search on graphs. To use the graph neural network as the controller, the parameters can be shared between controllers, thus greatly reducing the learning time of the controller. In [62], an evolutionary environment and a computational framework called Deep Evolutionary Reinforcement Learning are developed for exploring the relationship between embodied intelligence and the environment. This work also verifies the "Baldwin effect" in evolutionary biology through the morphological evolution learning mechanism, which tells that the behaviors learned by an organism will affect the species evolution. It is quite efficient as the evolutionary algorithm is essentially equivalent to a zero-order optimization.

1.3.5.2 Direct Reinforcement Learning Method

With the rapid development of reinforcement learning, some works have tried to directly use reinforcement learning to co-optimize the morphology and control [64]. Reference [165] jointly calculates morphology parameters together with control parameters using PPO (Fig. 1.10). It is difficult for learning because the morphology search space is large and the morphology and control are coupled. To this end, the author restricts the morphology search space by only optimizing the parameters of the specified robotic components, but does not the structure of the robot. On this basis, Ref. [164] uses reinforcement learning to realize morphology search and control policy learning for legged soft robots, and further realizes the sim-to-real transfer. Aiming at the self-assembling agents, Ref. [146] unifies the morphology and control in the action space. The morphological search and control design are described as a reinforcement learning problem, and the corresponding dynamic graph network controller is designed to make its morphology align with the robot morphology. Reference [64] uses reinforcement learning to jointly learn morphology and control policies for walking over obstacles. Reference [214] splits the morphology transformation and control optimization into two stages during the learning process, and uses the policy gradient method to jointly optimize the morphology and control actions. Recently, Ref. [116, 117] introduces the graph neural network into the reinforcement learning framework for morphology and control co-optimization, which provides a feasible way for the sim-to-real transfer in morphology learning.

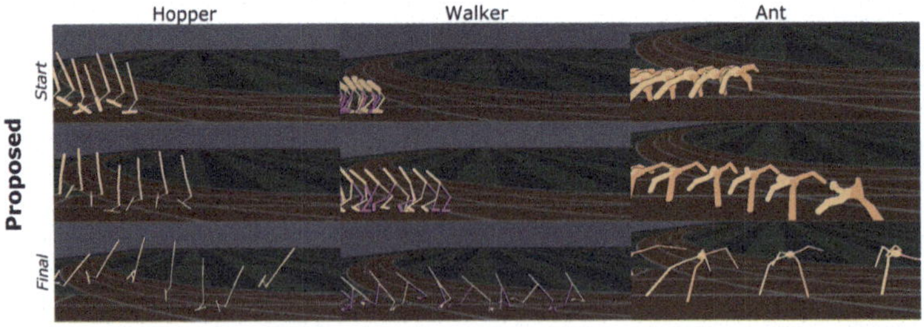

Fig. 1.10 Morphology-control collaborative optimization based on reinforcement learning [165]. (Reproduced with permission from Ref. [165])

1.3.5.3 Physical Implementation

Although the learning process of co-optimizing the morphology and control is usually carried out in a simulation environment, people have been trying to realize the learned morphology in the physical world. As early as 2000, Ref. [106] uses 3D printing technology to realize the morphology obtained from the evolution. However, it is also found that there is a big difference between the simulation and physical environment. The morphology with the satisfactory performance in the simulation can not achieve the desired performance after being physically produced [74]. To solve this problem, Ref. [18] proposes a continuous morphology modeling method. Reference [95] investigates how to incorporate the sim-to-real transfer effects into the optimization process as well. Reference [156] further investigates the relationship between this sim-to-real difference and the morphological complexity for a flapping wing, and it is shown that the sim-to-real gap is actually non-monotonic. For different types of robots, some works firstly perform morphology evolution in the simulation, and then using transferring techniques to implement them in the physical robots. Works such as legged robots [140, 160], soft robots [97], modular robots [6] mainly follow a sim-to-real transfer approach. With the increasingly various kinds and low cost of robot components, it is possible to directly optimize the agent morphology in physical systems. Reference [20] uses a robotic arm to manipulate different cubic modules, and achieves the evolutionary optimization by encoding the module and physically evaluating the performance of the assembled morphology (Fig. 1.11). More recently, Ref. [138] designs a quadruped robot with linear actuator mounted on its femur and tibia, allowing their lengths to be changed. Based on these optimizations, a learning approach for direct physical morphology evolution is obtained and experiments are carried out on different terrains. Overall, research on morphological optimization that is directly applied on physical systems is still at a beginning stage, and parameters that can be optimized are very limited. So far, the morphology optimization has been relatively successfully applied to the structure design of soft robots and robotic hands.

The development of robot morphological evolution is relative slow over the past years. An important reason is that people have only focused on a limited combination of rigid elements. Regarding to this problem, Ref. [70] utilizes a genetic algorithm to realize the morphology optimization of a voxelized soft robot including multiple material properties. Reference [34] studies the morphology of

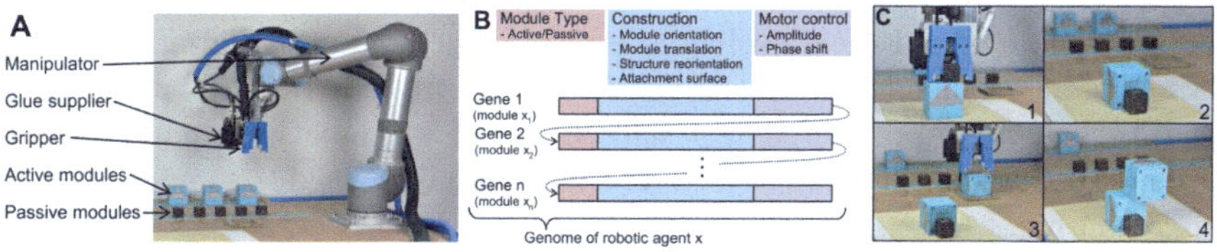

Fig. 1.11 Robotic arm devices which directly conduct morphology evolving with physical robotic system [20]

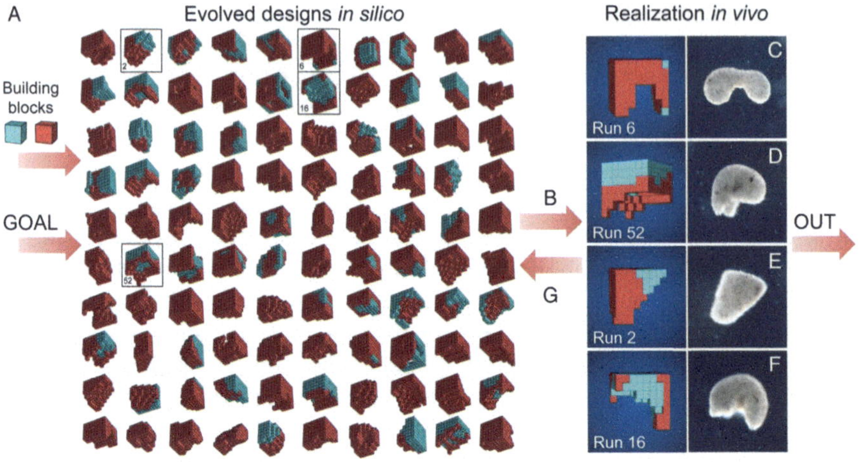

Fig. 1.12 Biological organic evolution [96]

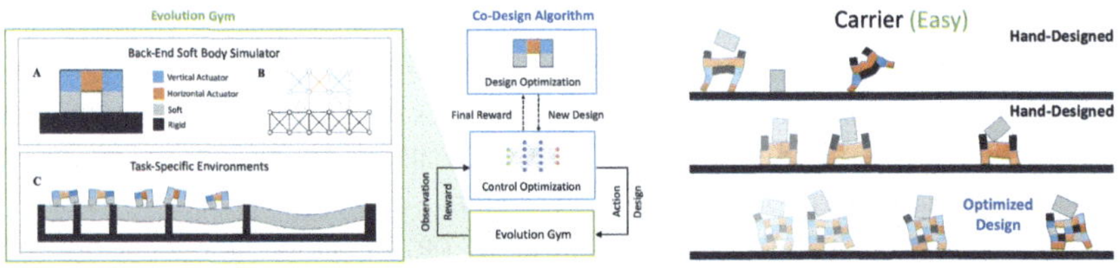

Fig. 1.13 The morphology optimization of voxel-based soft robots. Collaborative evolving framework developed in Ref. [11] and VSR in Ref. [11]

voxel models of different materials, which roughly simulate bones, tissues, muscles, etc. Reference [33] further embeds the control system into the physical simulation of the robot morphology, and proposes the so-called evolved electrophysiological soft robots. Recently, Ref. [96] uses direct encoding method to realize the morphological evolution of 3D voxel soft robots and it is applied to organisms realization (Fig. 1.12). As the morphological optimization, control, and production of 3D voxel soft robots are very difficult, some works begin to transfer to the evolutionary learning of 2D voxel soft robots [125]. Reference [87] also studies the phenomenon of "metamorphosis" in soft robots. Most of the work mainly considers the morphological evolution itself, and the control optimization during the environmental interaction is less considered. To this end, Ref. [11] proposes to realize the co-design of morphology and control of soft robots by combining the evolutionary optimization with reinforcement learning techniques. For 2D voxel-based soft robot (VSR), an Evolution Gym environment is developed (Fig. 1.13). The robots in the Evolution Gym are composed of many "cells" as basic units, including soft cells, rigid cells, and actuator cells that can actively contract or

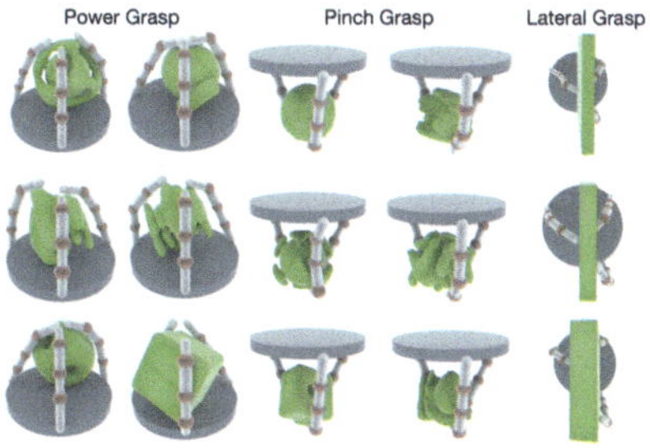

Fig. 1.14 The optimization for robotic hand morphology. Baysian optimization method [144]. (Reproduced with permission from Ref. [144])

expand. This allows the robot to freely evolves in its shape, and finally accomplish a series of tasks such as moving and manipulating objects on different terrains.

Another typical co-optimization of morphology and control application is the robotic hand morphology optimization. The optimized morphology should be able to adapt to objects that need to be manipulated and grasped. Reference [1] applies evolutionary strategies to optimize the morphology of robotic gripper, and graph element networks are introduced to improve the search efficiency. And the fine-tuning of the morphology for the customized two-finger gripper is realized (Fig. 1.14 LEFT). Aiming at the power grasp, pinch grasp and lateral grasp, Ref. [144] projects the morphology and control parameters of the hand into the common latent space, and proposes a Bayesian optimization algorithm to search for the best hand morphology (Fig. 1.14 RIGHT). Recently, Ref. [206] designs a general morphology representation based on the cage-based deformation models from a perspective of computer graphics. The advantage of this method is that rich morphology can be described with fewer parameters. By combining this model with a differentiable simulator, an end-to-end learning approach is formed.

1.3.5.4 Summary

Most of the current research work is verified in the simulation environment. How to transfer the morphological evolution results in the simulator to the physical system, or directly evolve in the physical system, is a frontier research direction worth further discussion in the future. In addition, it is worth noting that there are already many mature approaches for the morphology design in the field of mechanisms. How to introduce these empirical information, as well as related physical constraints and external knowledge, in combination with data-driven learning methods to co-optimize the morphology and control will be a helpful tool to improve learning efficiency.

1.3.6 Perception-Driven Action Generation ($\mathcal{P} \rightarrow \mathcal{A}$)

This part corresponds to the link ⑤ from *Perception* to *Action* in Fig. 1.1. The environmental perception information is used to guide the agent to generate actions to interact with the environment in real time. This paradigm is most commonly employed in many robotic tasks. For example, in the field of autonomous driving, vision perception and other senses are used to for autonomous navigation. Many robotic manipulation tasks also resort to visual perception to generate grasp synthesis. Most of the current machine perception studies also fall into this category.

1.3.6.1 Traditional Visual Manipulation and Navigation

During the past decades, sensors and perception algorithms are rapidly developing. Many latest hardware and algorithms have been integrated into robot platforms. The typical commonly used sensors includes Kinect, Realsense, Event Camera, etc. And the computer vision algorithms [211, 226] for traditional tasks such as object detection, recognition, segmentation, and tracking are further improved in the deep learning era. Under such circumstance, the perception-action tasks have been greatly developed, in which robotic manipulation and visual navigation are two representative tasks.

For the robotic manipulation task, early works mainly fall into the category of analytical approaches [13], which use the contact models to calculate proper contact forces and torques. However, the analytical approaches require full knowledge of the object which is difficult to obtain with only partial and noisy perception information. On the other hand, with the big success of deep learning techniques in computer vision, more and more data-driven approaches have been proposed for robotic manipulation [135]. And end-to-end methods in which a single network regresses actions directly from the visual input become more and more prevalent. In the visual navigation task, the agent is supposed to plan a reasonable path to the destination. With the help of visual simultaneous localization and mapping (SLAM), the agent is able to well perceive and understand its environmental surroundings [22]. Recently, more advanced SLAM algorithms such as metric-semantic SLAM [166] and dynamic SLAM [195] are proposed to handle with dynamic environments. Traditional robotic visual navigation usually follows a multi-stage paradigm of global path planning combined with local motion control, which requires much prior knowledge and engineering design effort. Similar to the robotic manipulation, more and more end-to-end learning approaches have been proposed, which directly map visual inputs to motion commands [205]. Beyond question, perception-driven action generation is extremely common in the field of robotics and we only focus two important emerging directions including multi-modal fusion and language guidance, which are also can be combined into a unified language-conditioned multi-modal perception-driven action generation framework [204] (Fig. 1.15).

1.3.6.2 Multi-modal Perception-Driven Navigation and Manipulation

Multi-modal perception is extensively used for navigation. Generally speaking, visual and depth information are necessary for mobile robots. Kinect, Realsense,

Fig. 1.15 Language-conditioned perception-driven action generation [204]

ultrasonic, and laser radar are often used in the indoor or outdoor scenarios [49]. A series of multi-modal datasets have been developed for research and industrial purposes [55, 217]. Recently, other perception modalities such as 4D millimeter wave radar also find potential applications in some extreme weather conditions [191]. On the other hand, multi-modal perception is also vital for some tasks that could be difficult to be solved with single perception modality. For example, visual, tactile and auditory modalities are often fused together to recognize different attributes of object. Reference [14] reviews the progress made in integrating multi-modal perception in robotic manipulation. Reference [110] systematically addresses the weakly-pairing problem in fusion of visual-auditory-tactile modalities for material recognition. Recently, the multi-modal perception has been used for more extensive and complicated tasks such as precise grasp [194] and in-hand manipulation [176].

1.3.6.3 Language-Conditioned Navigation and Manipulation

Another important direction is to ground the language to the perception and therefore it is possible to provide natural interfaces with human. A representative task is the Visual Language Navigation (VLN) task, in which the agent carries out the navigation instruction by aligning the language with its visual perception and the generated navigation actions [52]. The agent needs to simultaneously understand the navigation instructions and the visual perception information, and then corresponding navigation actions are generated. Reference [3] proposes a basic framework for the VLN problem and a benchmark is established in the MatterPort3D environment. The current research on visual language navigation task involves a lot of latest methods in the computer vision, natural language processing and cross-modal processing to help agents better solve the task. Along the success of VLN, many works extend this idea to the manipulation scenarios and form the so-called Visual-Language-Manipulation (VLM) task [220]. In essence, VLM exhibits similar characteristics as VLN but may be faced with more challenges because the manipulation usually requires more fine-grained action space and more complicated planning.

1.3.6.4 Summary

The perception-driven action generation tasks are very common in robotics. With the rapid development of deep learning and computing techniques, end-to-end model that directly projects the perception information to action space have become the mainstream approaches to solve this class of problems. Furthermore, the multi-modal perception and language-conditioned action tasks have become more and more popular in practical scenarios. However, there still exists great challenges to align the natural language, multi-modal perception and complicated action together.

1.3.7 Action-Driven Perception Improvement ($\mathcal{A} \rightarrow \mathcal{P}$)

This part corresponds to the link ⑥ from *Action* to *Perception* in Fig. 1.1, which emphasizes that the perception can be improved with goal-oriented exploration. Research work in this area has attracted attention as early as the 1980s, when the active perception is first systematically discussed [7]. In the active perception, the agent takes motions under intelligent control to improve how it perceives the environment (Fig. 1.16). It is theoretically proved that after the introduction of motion, many ill-posed or nonlinear problems in the field of traditional computer vision become well-posed and linear problems [2]. Since then, active perception has been widely used in robotics and becomes an essential part of embodied intelligence [186].

In the early days, the information theory methods act as the typical method for the active perception. For example, Ref. [44] attributes many visual problems to state estimation problem, and uses information theory methods to study the optimal sensor selection for state estimation, which lays the foundation of Bayesian approaches for active perception [147]. Most of these methods use conditional entropy or mutual information to measure uncertainty, and usually demonstrates poor generalization capability. In recent years, reinforcement learning has gradually become an important approach to achieve active perception [143]. In the following, we summarize several typical active perception tasks, and point out the recent trends of incorporating semantics and language into active perception tasks.

Fig. 1.16 Active perception exhibits typical *Time for Space* characteristics

1.3.7.1 Active Visual Perception

Any typical computer vision task (e.g. detection, recognition, etc.) can be transformed to a corresponding "active" version by introducing the action. Here we list some typical active visual perception tasks. One is the active object detection [66], in which the object to be detected is not in the initial field of view, and the agent needs to explore the environment in order to discover the object. Or the pose of the camera is adjusted by some intelligent control strategies in order to better detect object that is occluded or in small scale. For the active object recognition task [83], it aims at adjusting the camera pose to obtain a new field of view when the feature extracted in the current field of view is not discriminative enough for the recognition. For example, to recognize a person, it is much more easier to recognize a person in the front rather than from the back of the head. There is also a kind of active object tracking task [222]. The traditional visual object tracking task is required to accurately detect and temporally correlate the objects in the video streaming. When an object moves out of the field of view, the tracking algorithm can no longer function well. To this end, the active object tracking can adjust the camera pose in real time to keep the moving object always in the field of view.

1.3.7.2 Visual Semantic Navigation

In recent years, there is an important trend to investigate the language-conditioned active perception. The most typical task in this respect is the Visual Semantic Navigation (VSN). Unlike VLN, which uses a long language navigation instruction, VSN only requires the user to provide a semantic target object, and then the agent can autonomously search it in the environment. Such semantic targets can be given by semantic labels [210], or object images [225] (Fig. 1.17). VSN tasks in fact are the extension of active object detection and recognition tasks. As the object to be detected in the VSN task is described with semantic information, it provides the possibility to interact with people.

Currently, the VSN models are mostly trained in simulation environments, which brings the gap between the simulation and real-world environment. Reference [113]

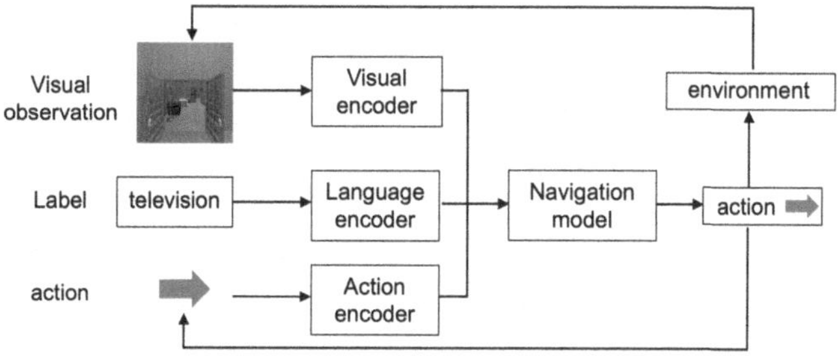

Fig. 1.17 The typical frameworks for VSN

discusses the embodied mismatch problem where the actual action space does not match that in the simulation environment. Reference [216] utilizes the robot's own capabilities, combining with the 3D perception information to improve its object discovery ability. Reference [56] constructs an actual physical scene to comprehensively evaluate the performance difference of different methods in the simulation and physical environment. And it has pointed out that the modular design method significantly outperforms the end-to-end learning method.

On the other hand, as the prior information of the scene will often bring great convenience to the discovery of objects, there are methods leveraging the scene prior for efficient object discovery [210]. However, these works ignore an important problem. As time goes by, the scene itself may experience various unexpected changes. Neglecting the dynamic nature of the scene will severely deteriorate the performance of object discovery ability, and it also brings difficulties for a long-term deployment. Reference [224] establishes a dynamic spatial-temporal scene graph to solve this problem, and the scene graph is updated in real time with the information collected by the agent during its daily explorations so as to dynamically adapt to changes in the environment. Reference [99] formulates the dynamic scene graph updating as a link prediction problem, and designs a dynamic simulator to generate dynamic scenes. Reference [105] proposes to dynamically obtain and predict the people's locations in real time by monitoring the dialogue. Generally speaking, the processing of dynamic scenes is still at a very infancy stage. Currently, the active perception tasks develop rapidly. Many tasks such as rearrangement [197], A-V navigation [30, 112], and tidy-up [162] are newly proposed. Various perception modalities, such as vision, hearing and touch, are also employed in the active perception tasks [17, 111].

1.3.7.3 Embodied Question Answering

The Embodied Question Answering (EQA) task [39, 59] is a more advanced topic in the field of active perception. In the EQA task, the agent starts from a random position in the 3D environment, and it actively explores the environment to answer a given natural language question (Fig. 1.18). However, the development of current EQA task still faces many difficulties in language understanding, task planning, visual perception and action execution.

First, the original EQA task designs the answer to the question beforehand, so it is actually a multi-classification problem. The answers lack the ability to describe the surrounding environment. Inspired by technologies such as image captioning, Ref. [179] studies the problem of embodied scene description, where the agent actively explores the environment to yield information-rich semantic descriptions for the scenes. It can effectively support the environment description ability of the embodied question answering system. Secondly, current problems in the EQA task are very simple and straightforward. To address the connection between the complex logical reasoning and physical world, Ref. [212] proposes Multi-Target Embodied Question Answering, which extends the original EQA task from a single-

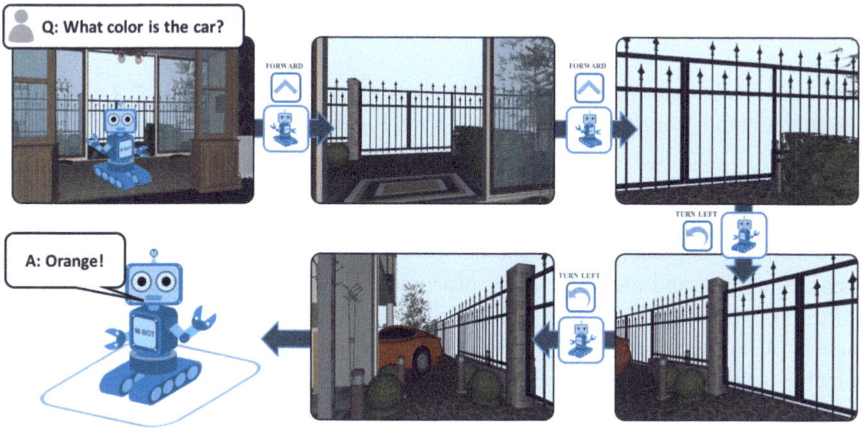

Fig. 1.18 Embodied question answering task [39]. (Reproduced with permission from Ref. [39])

target setting to a more challenging multi-target setting. Reference [178] further uses the knowledge graph to realize the understanding and reasoning of complex questions. Thirdly, currently EQA work exhibits insufficient scene generalization ability. At present, it is generally improved by means of visual feature extraction and multi-modal perception. For example, Ref. [200] introduces more types of input modalities such as RGB-D and point cloud to embodied question answering tasks. Reference [119] improves the performance of embodied question answering task by introducing video segmentation and corresponding attention mechanism. Additionally, Ref. [40] decomposes the embodied question answering process into multiple steps through a neural module network, and it is found that this modular approach significantly improves the performance compared to the end-to-end learning method. In order to improve the efficiency of the environment exploration, Ref. [180] proposes a multi-agent EQA, which uses the collaboration of multiple agents to accomplish the EQA task.

1.3.7.4 Summary

Active perception is a class of typical embodied intelligence task, in which the perception is improved while the agent taking actions. It presents a *Time for Space* characteristics (Fig. 1.16), which is actually a double-edged sword. Therefore, its application is still rather limited due to its complexity and difficulty for deployment. Recently, semantics and language are more and more incorporated in the active perception, which may provide more chance for the wide applications of active perception. However, the active perception for dynamic scenes is still very challenging.

1.3.8 Action-Driven Embodied Learning ($\mathcal{A} \rightarrow \mathcal{L}$)

This part corresponds to the link ⑦ from *Action* to *Learning* in Fig. 1.1. Learning is the fundamental component of embodied intelligence, and only when an agent has the ability to autonomously update its own knowledge and skills, can it be able to adapt to the ever-changing open environment. Currently, various large-scale datasets are constructed from abundant Internet data, and the pretrained perception models have achieved unprecedented success. However, under such learning paradigm, the data acquisition and model learning processes are separated. Whether a dataset is truly suitable for some learning task is still an open problem. In addition, operations such as cleaning data and labeling data must be performed before training the model, which is far from the requirements of autonomous learning.

To leverage the action of the embodied agent to improve perception performance is actually the way how humans learn. For example, when humans see a new object, we will naturally manipulate or look the object from various viewpoints to actively explore it, and we can then realize that no matter from which viewpoint we look at the object, it is the same object. We learn continuously in this action-learning loop. The exploration process is actually a data collecting and labeling process. Inspired by this, the embodied agent is encouraged to explore areas where the pre-trained perception model may not perform well. Data observed during the exploration can then be collected and labeled for further improving the perception model. This action-driven embodied learning paradigm usually can be divided into three stages: exploration policy learning, training samples collection, and application deployment. It combines the data acquisition and model learning seamlessly, and can continuously improve the agents' capability during the exploration. It has already become a very promising direction in the embodied intelligence. The main achievements can be classified as learning for perception, and learning for representation, which will be detailed as follows.

1.3.8.1 Learn to Improve Perception

Reference [28] introduces a semantic curiosity reward, which uses the consistency of the semantic map of the current frame and the previous frame to guide the agent to explore the environment and collect data that is beneficial to improve model performance (Fig. 1.19). Reference [26] constructs a 3D global scene semantic map, which can be back-projected to the 2D space to generate pseudo-labels, thereby realizing the self-supervised learning. Meanwhile, the number of voxels in 3D semantic voxels is used as a reward to encourage the agent to explore locations with more semantic information. Reference [190] analyzes the different roles of familiar and unfamiliar samples in the process of embodied learning, and designs a new self-supervised learning method. All of the above works are realized in the

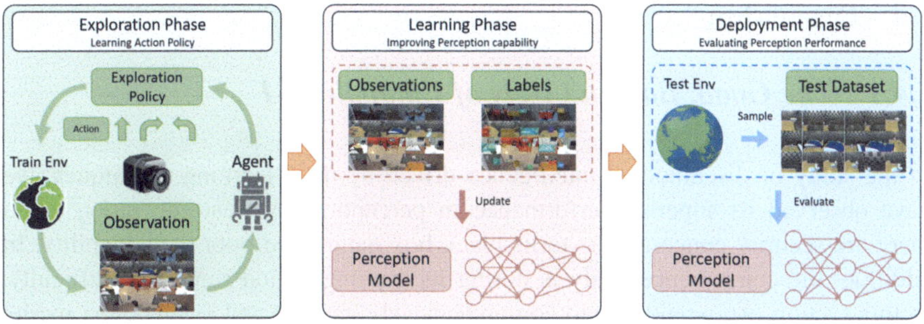

Fig. 1.19 A semantic curiosity reward is designed to learn an exploration policy [28]

simulation environment. Further, Ref. [86] learns exploration trajectories based on the uncertainty of the semantic distribution and tries to realize the physical implementation. Reference [219] proposes a 3D embodied dataset in real scenarios, which includes the visual data and the dense point cloud data. It uses the consistency of 3D perception results under different views to collect the data.

Most of these current works are aimed at some general perception models such as object detection, object segmentation, etc. The embodied learning can also be combined with some specific tasks to enable some goal-oriented tasks. For example, Ref. [103] proposes a general method for the agent to automatically learn to construct the scene graph. Reference [122] proposes the embodied NeRF, which aims to automatically collect good images to construct 3D scenes.

In addition to directly improving the performance of perception models, there is another type of research that directly conducts feature representation learning through autonomous behavior learning [198]. Reference [21] explores curiosity-driven representation learning from a set of games. Reference [154] utilizes a robotic arm to physically interact with objects using simple tactile sensors to help learn visual features and promising results are obtained. Reference [207] further proposes DensePhysNet, which learns several physical properties of objects by actively performing a series of dynamic interactions.

1.3.8.2 Summary

It can be observed that the action-driven embodied learning can provide a learning mechanism which can be further used for some down-streaming tasks, such as perception or understanding. Its core is to leverage the action of the embodied agent for learning, which bridges the gap between the data collection and model training. In this way, the embodiment characteristic of the agent is fully leveraged.

1.4 Discussion

1.4.1 The Good, Bad and Ugly of Embodied AI

In the study of disembodied intelligence driven by deep learning techniques, we have observed its superior performance in perception and learning tasks, while there are security concerns due to its black box nature and poor explainability. In addition, the convergence analysis during its training period remains a difficulty. In this section, we would like to comprehensively analyze and evaluate the merits (Good), demerits (Bad) and difficulty (Ugly) in the embodied intelligence.

1.4.1.1 Good

The most important characteristic of embodied intelligence is that it introduces the "interaction between the body and the environment", and thus the action dimension is introduced. Many difficult tasks can be easier if the body is used wisely. For example, it is difficult to recognize an person when observing from the back. To solve this problem, the typical approach in disembodied intelligence is to extract better features using prior knowledge or data-driven methods, while the typical approach in embodied intelligence is to move to a better viewpoint so as to make the recognition task easier. On the other hand, the appropriate use of body morphological information can also significantly simplify the design of the controller. For example, properly designing the aerodynamic configuration of an aircraft can leverage the environment to achieve more flexible movements, whereas it will be very difficult to solely rely on the design of control algorithms.

1.4.1.2 Bad

The traditional embodied intelligence overemphasizes the effect of body and environment, while it ignores the role of knowledge and data. In this way, the development of the brain and body is not balanced. On the other hand, the security of embodied intelligence is also a very challenging problem due to the physical interaction with the environment. For example, in the task of object recognition, if the model has achieved a recognition accuracy of 95%, it can be deployed in some practical applications, while in an embodied intelligence system, even if a 1% failure rate (such as in a surgery) may result in fatal result.

1.4.1.3 Ugly

Embodied intelligence emphasizes that the intelligence emerges from the tight coupling of the environment, body and brain. This process requires an integration of morphology, perception, action, learning, and even language understanding. Since the morphology, perception, action, learning and language belong to different fields and have significantly different characteristics, it is very challenging to achieve the alignment of all of them. On the other hand, the employment of embodied intelligence requires the perfect realization of various components, such as environmental perception, language understanding, behavior control, etc. Therefore, it is difficult to be realized in an elegant end-to-end manner.

1.4.2 Embodied Intelligence and Robotics

The embodied intelligence is closely related to robotics, and they are even confused with each other in many situations. The core of embodied intelligence is that the intelligence generated from the interaction between the body and environment. The body here is not only limited to a robot. In fact, any physical body (such as the famous centrifugal governor invented by *James Watt*) or non-physical body (such as digital human and simulation agent) are applicable. Therefore, the embodied intelligence focuses more on the study of general theories, methods and technologies, and robots can be regarded as an embodiment of the embodied intelligence. For the robot itself, its intelligent behavior may either come from the embodiment intelligence or not.

On the other hand, we can interpret the term "embodied intelligence" as using "embodied" to describe "intelligence". The "intelligence" generally could refer to intelligent tasks such as visual detection, language question answering, which belong to the category of disembodied intelligence. But by including "embodied" in front, we have new tasks such as embodied visual detection, embodied question answering, etc. In this sense, the embodied intelligence can be understood as an extension of disembodied intelligence, and it is meant to move towards the robotics field. On the other side, in the field of robotics, since robots generally have physical bodies, people instead seldom use "embodied" to modify the related research work in the early days. Currently, the robotics and embodied intelligence areas are indeed deeply integrated.

1.5 Frontiers of Embodied Intelligence

1.5.1 Inspiration from Disembodied Machine Learning

The supervised learning method that is advocated in the traditional machine learning field requires a clear separation between training set and test set, and accurate labels should be provided. It is obvious that it will hinder the agent's adaptation to dynamic and open environment. Fortunately, the machine learning field is also aware of this problem and great efforts have been made to solve it. Self-supervised learning provides a new type of learning paradigm to alleviate the labeling pressure. Constructing pseudo-labels for unlabeled samples by various means can achieve good results in the feature learning stage [85]. On the other hand, lifelong learning, although originated in the field of robotics [183], has also received great attention in the field of machine learning in recent years, and various methods to overcome catastrophic forgetting have been proposed in tasks such as recognition, detection, etc. [175]. It has been extended to the field of reinforcement learning as well, which has brought great opportunities for embodied agents to improve their performance in the real-world environment. Inspired from these achievements in the disembodied machine learning, we believe that the seamless combination of self-supervised learning and lifelong learning could also provide a promising way for the embodied intelligence.

1.5.2 The Roles of Morphology

Many embodied navigation tasks that are currently studied in the simulation environment usually only use a simple action space, and the morphological characteristics of the agent are not considered [98]. Besides, there are high constraints on the environment, in which a traversable topological map is often required, and issues such as obstacle avoidance are not considered. As mentioned above, in the embodied intelligence, morphology is an important tool to generate intelligence and an important way to reflect intelligence. It is regarded as the core of embodied intelligence. However, works such as morphological computation, morphological control, and morphological optimization have not become important research directions that support the development of embodied intelligence. An important reason is that the success of morphology is strongly dependent of the material, which still exhibits great challenges in selection, actuation and design [38]. Recently, the soft material has found extensive applications in the embodied intelligence because intrinsic properties of materials in soft robots allow for reducing the mechanical and algorithmic complexity in ways not possible with rigid-bodied robots. This provides great research space for the future work.

1.5.3 Sim2Real

With the rapid development of computer graphics, the visual simulation technology can provide a realistic platform for the research of embodied intelligence, and lays a good foundation for the establishment of standardized benchmarks [15]. At present, there are many simulation platforms directly applied to the embodied intelligence tasks such as AI2THOR [42], Matterport3D [25], Habitat [163] and iGibson [168]. However, although the visual effect in the current simulation environment is more and more realistic, it is impossible for the simulation environment to cover all aspects in the real-world. In addition, the perception information is mostly restricted to the vision modality in the current simulation environment. There still exist great difficulties to simulate audio, tactile and other perception modalities. On the other hand, when we transfer the control strategy learned in the simulation to the real-world, we will face the great behavior gap since it is difficult to model the dynamics of the embodied agent. Therefore, the embodied agents trained in the simulation environment still face serious transferring problems when they directly transfer to the physical world, which is called the Sim2Real problem. Currently, there are mainly two ways to solve this problem. One way is to use various means to enrich the simulation environment, such as domain randomization [128], and the other is to narrow down the perception and action space gap between the simulation and the physical environments as much as possible. For example, Ref. [27, 114] proposes the semantic perception approach. Reference [120] proposes an environment augmentation method to increase the diversity of training simulation environments. Reference [113] considers the embodied mismatch of action space. Reference [114] establishes a Sim2Real method to align the simulation and physical worlds using semantics and high-level planning. Reference [221] constructs a structure-aware representation for generalizable active object tracking. Reference [227] involves both virtual subset and real subset for training.

1.5.4 Multi-modal Embodied AI and Language Grounding

"Multi-modal" is frequently mentioned in the current embodied intelligence research [101]. The recently released VIMA is known as a multi-modal artificial intelligence system [84]. But the "Multi-modal" here generally refers to the vision, language, sound, etc, and it is more focused on solving the relationship between the visual environmental perception and human instructions and intentions. Essentially, the multi-modal fusion is still limited in the domain of disembodied intelligence.

Embodied intelligence emphasizes the interaction between the body and the environment. This interaction includes a variety of different modalities such as vision, hearing, and touch. How to effectively fuse these heterogeneous multi-modal information to realize multi-modal embodied perception is an important issue that needs to be paid attention [194]. Currently, vision and hearing are relatively more investigated areas [29, 54, 189], and the corresponding sensors are developing rapidly. The study of tactile sensing relatively lags behind, which seriously restricts the development of embodied intelligence, while it is important for the physical interaction between the body and the environment [109]. It can be used to ensure the safety, stability and flexibility of the operation, and it can obtain unique information that is difficult to capture by other sensing modalities.

The field of language grounding is also a key area of embodied AI and robotics [60]. Grounding refers to the capability of the agent to intrinsically connect its own internal linguistic representation and symbols with the referents in the external world, as well as internal states. Thus an embodied robot must learn to associate for example the words "red" and "ball" with an external toy, the ball, and its perceptual properties such as the red color. But grounding does not have to necessarily link an internal symbol with an external object in a direct manner. In fact, even abstract words can be indirectly grounded in sensorimotor experience and internal feelings [24].

1.5.5 Large Language Model for Embodied AI

Recently, the Large Language Model (LLM) has received great attention. Especially, the emergence of GPT and other LLMs have brought many new paradigms for robotic applications. Reference [84] develops the multi-modal prompt for manipulation tasks. Reference [104] investigates how pre-trained visual-language models can benefit the language-conditioned robotic manipulation. References [215] and [43] attempt to use LLM to solve the task assignment in the multi-robot collaboration problem and promising results are demonstrated. Reference [90] shows some interesting applications of AI-generated content (AIGC) in robotic manipulation. Very recently, Ref. [141] assembles a dataset from 22 different robots collected through a collaboration between 21 institutions, demonstrating 527 skills. All of these attempts show the strong capability of LLM in robotic applications. However, we must also see that though the introduction of these technologies may bring new ideas and opportunities for the application of embodied intelligence, many key challenges of the embodied intelligence have not been really solved. In particular, LLM are better at solving high-level task planning and human-computer language interaction problems, and there is still no ideal solution for low-level control in navigation and manipulation tasks [209].

1.5.6 Multi-agent Collaboration

The multi-agent collaboration is an indispensable part in the embodied intelligence. Different from some works that learn multi-agent collaboration strategies in simple disembodied environments [63, 79], there emerge a bunch of studies that learn multi-agent collaboration strategies in the embodied visual environments, and the multi-agent reinforcement learning has become the mainstream method to learn the multi-agent collaboration strategy [32, 35]. In [81, 82], two agents learn to collaboratively lift and move furniture in the action level. Some typical embodied tasks such as embodied question answering and embodied navigation have also been extended to the multi-agent setting [181, 188]. However, most of multi-agent collaboration tasks only consider the homogeneous agents, and there are already works trying to leverage different capabilities of heterogeneous agents to deal with more complex tasks [127]. Besides, the multi-agent collaboration also brings the challenges in the communication mechanism among agents. Reference [145] analyzes different communication mechanisms between heterogeneous agents. Reference [92] incorporate the communication information in the multi-agent reinforcement learning. It is worth noting that besides multi-agent collaboration, it has been demonstrated that the competition among multiple agents could also facilitate the skill learning ability of the agent [76, 203, 223]. With the increasingly complex tasks, it is clear that the multi-agent collaboration will become more and more important and necessary [50].

1.5.7 Security and Vulnerability of Embodied AI

The security issues of artificial intelligence technology represented by deep learning has attracted more and more attention. Some pixel-level attacks will lead to the recognition failure or even decision-making failure of the entire intelligent system. For the embodied intelligence system, as it involves physical action execution and interaction, its security problem should be even more emphasized. On one hand, we can see that the performance of many existing embodied perception tasks drops significantly when faced with unseen scenes or objects, and the transferring of policies trained in the simulation environment to the physical environment also faces huge challenges. On the other hand, even for some successful application scenarios, the performance of the embodied exploration can be affected by some attacks. For example, Ref. [107] directly reduces the accuracy of embodied question answering by introducing texture effects on the environment. Reference [218] studies the Byzantine-robust federated embedded agent learning for the VLN task. Aiming at audio-visual (A-V) navigation, the attack of sound source can also fail the navigation task [213]. In the context of embodied intelligence, the more ways for attack, the more ways for defense will also emerge. This will be an important topic for a long period of time in the future.

1.6 Summary

In this chapter, we review the development of embodied intelligence, especially the embodied artificial intelligence. From the perspectives of morphology, action, or perception and learning, a unified embodied intelligence framework is proposed, which mainly focuses on connections between these components rather than themselves alone. According to the proposed framework, a thorough summary and categorization of existing studies are provided. Additionally, we also identify areas where future research can benefit from these intrinsic connections. As the embodied intelligence is a synergy of morphology, action, or perception and learning, the community is also faced with both great opportunities and challenges that requires an integration of technologies from multiple different disciplines.

References

1. Alet F, Bauza M, Adarsh KJ, Thomsen M, Rodriguez A, Kaelbling LP, Lozano-Pérez T (2020) Robotic gripper design with evolutionary strategies and graph element networks. In: NeurIPS Workshop on Machine Learning for Engineering Modeling, Simulation, and Design, vol 23
2. Aloimonos J, Weiss I, Bandyopadhyay A (1988) Active vision. Int J Comput Vis 1:333–356
3. Anderson P, Wu Q, Teney D, Bruce J, Johnson M, Sünderhauf N, Reid I, Gould S, Van Den Hengel A (2018) Vision-and-language navigation: Interpreting visually-grounded navigation instructions in real environments. In: Proceedings of the IEEE Conference on Computer Vision and Pattern Recognition, pp 3674–3683
4. Auerbach JE, Bongard JC (2010) Dynamic resolution in the co-evolution of morphology and control. In: Artificial Life XII: Proceedings of the Twelfth International Conference on the Synthesis and Simulation of Living Systems, CONF, pp 451–458. MIT Press, Cambridge, MA
5. Auerbach JE, Bongard, JC (2012) On the relationship between environmental and morphological complexity in evolved robots. In: Proceedings of the 14th Annual Conference on Genetic and Evolutionary Computation, pp 521–528
6. Auerbach JE, Concordel A, Kornatowski PM, Floreano D (2018) Inquiry-based learning with robogen: An open-source software and hardware platform for robotics and artificial intelligence. IEEE Trans Learn Technol 12(3):356–369
7. Bajcsy R (1988) Active perception. Proc IEEE 76(8):966–1005
8. Barrett L (2011) Beyond the brain: How body and environment shape animal and human minds. Princeton University Press, Princeton
9. Bellman R (1957) A Markovian decision process. J Math Mech, 679–684
10. Bernth JE, Ho VA, Liu H (2018) Morphological computation in haptic sensation and interaction: from nature to robotics. Adv Robot 32(7):340–362
11. Bhatia J, Jackson H, Tian Y, Xu J, Matusik W (2021) Evolution gym: A large-scale benchmark for evolving soft robots. Adv Neural Inf Proc Syst 34:2201–2214
12. Bhovad P, Li S (2021) Physical reservoir computing with origami and its application to robotic crawling. Sci Rep 11(1):13002
13. Bicchi A, Kumar V (2000) Robotic grasping and contact: A review. In: Proceedings 2000 ICRA. Millennium conference. IEEE international conference on robotics and automation. Symposia Proceedings (Cat. No. 00CH37065), vol 1, pp 348–353. IEEE
14. Billard A, Kragic D (2019) Trends and challenges in robot manipulation. Science 364(6446):eaat8414
15. Biswas A, Wang A, Silvera G, Steinfeld A, Admoni H (2022) Socnavbench: A grounded simulation testing framework for evaluating social navigation. ACM Trans Human Robot Interact (THRI) 11(3):1–24
16. Blake C, Kurin V, Igl M, Whiteson S (2021) Snowflake: Scaling gnns to high-dimensional continuous control via parameter freezing. Adv Neural Inf Proc Syst 34:23983–23992
17. Bohg J, Hausman K, Sankaran B, Brock O, Kragic D, Schaal S, Sukhatme GS (2017) Interactive perception: Leveraging action in perception and perception in action. IEEE Trans Robot 33(6):1273–1291
18. Bongard J, Zykov V, Lipson H (2006) Resilient machines through continuous self-modeling. Science 314(5802):1118–1121
19. Bongard JC, Pfeifer R (2003) Evolving complete agents using artificial ontogeny. In: Morpho-functional machines: the new species: designing embodied intelligence, pp 237–258. Springer, New York
20. Brodbeck L, Hauser S, Iida F (2015) Morphological evolution of physical robots through model-free phenotype development. PloS One 10(6):e0128444
21. Burda Y, Edwards H, Pathak D, Storkey A, Darrell T, Efros AA (2018) Large-scale study of curiosity-driven learning. arXiv preprint arXiv:1808.04355

22. Cadena C, Carlone L, Carrillo H, Latif Y, Scaramuzza D, Neira J, Reid I, Leonard JJ (2016) Past, present, and future of simultaneous localization and mapping: Toward the robust-perception age. IEEE Trans Robot 32(6):1309–1332
23. Cangelosi A, Bongard J, Fischer MH, Nolfi S (2015) Embodied intelligence. Springer handbook of computational intelligence, pp 697–714
24. Cangelosi A, Stramandinoli F (2018) A review of abstract concept learning in embodied agents and robots. Phil Trans Roy Soc B Biol Sci 373(1752):20170131
25. Chang A, Dai A, Funkhouser T, Halber M, Niessner M, Savva M, Song S, Zeng A, Zhang Y (2017) Matterport3d: Learning from rgb-d data in indoor environments. arXiv preprint arXiv:1709.06158
26. Chaplot DS, Dalal M, Gupta S, Malik J, Salakhutdinov RR (2021) Seal: Self-supervised embodied active learning using exploration and 3d consistency. Adv Neural Inf Proc Syst 34:13086–13098
27. Chaplot DS, Gandhi DP, Gupta A, Salakhutdinov RR (2020) Object goal navigation using goal-oriented semantic exploration. Adv Neural Inf Proc Syst 33:4247–4258
28. Chaplot DS, Jiang H, Gupta S, Gupta A (2020) Semantic curiosity for active visual learning. In: Computer Vision–ECCV 2020: 16th European Conference, Glasgow, UK, August 23–28, 2020, Proceedings, Part VI 16, pp 309–326. Springer, New York
29. Chen C, Al-Halah Z, Grauman K (2021) Semantic audio-visual navigation. In: Proceedings of the IEEE/CVF Conference on Computer Vision and Pattern Recognition, pp 15516–15525
30. Chen C, Jain U, Schissler C, Gari SVA, Al-Halah Z, Ithapu VK, Robinson P, Grauman K (2020) Soundspaces: Audio-visual navigation in 3d environments. In: Computer Vision–ECCV 2020: 16th European Conference, Glasgow, UK, August 23–28, 2020, Proceedings, Part VI 16, pp 17–36. Springer, New York
31. Chen T, Murali A, Gupta A (2018) Hardware conditioned policies for multi-robot transfer learning. Adv Neural Inf Proc Syst 31. https://doi.org/10.48550/arXiv.1811.09864
32. Chen Y, Geng Y, Zhong F, Ji J, Jiang J, Lu Z, Dong H, Yang Y (2023) Bi-dexhands: Towards human-level bimanual dexterous manipulation. IEEE Trans Pattern Anal Mach Intell. https://doi.org/10.48550/arXiv.2206.08686
33. Cheney N, Clune J, Lipson H (2014) Evolved electrophysiological soft robots. In: Artificial Life Conference Proceedings, pp 222–229. MIT Press, Cambridge, MA
34. Cheney N, MacCurdy R, Clune J, Lipson H (2014) Unshackling evolution: evolving soft robots with multiple materials and a powerful generative encoding. ACM SIGEVOlution 7(1):11–23
35. Ci H, Liu M, Pan X, Zhong F, Wang Y (2023) Proactive multi-camera collaboration for 3d human pose estimation. arXiv preprint arXiv:2303.03767
36. Collins S, Ruina A, Tedrake R, Wisse M (2005) Efficient bipedal robots based on passive-dynamic walkers. Science 307(5712):1082–1085
37. Corucci F, Cheney N, Lipson H, Laschi C, Bongard J (2016) Material properties affect evolutions ability to exploit morphological computation in growing soft-bodied creatures (jul 2016)
38. Coyle S, Majidi C, LeDuc P, Hsia KJ (2018) Bio-inspired soft robotics: Material selection, actuation, and design. Extreme Mech Lett 22:51–59
39. Das A, Datta S, Gkioxari G, Lee S, Parikh D, Batra D (2018) Embodied question answering. In: Proceedings of the IEEE Conference On Computer Vision and Pattern Recognition, pp 1–10
40. Das A, Gkioxari G, Lee S, Parikh D, Batra D (2018) Neural modular control for embodied question answering. In: Conference on Robot Learning, pp 53–62. PMLR
41. Degrave J, Caluwaerts K, Dambre J, Wyffels F (2015) Developing an embodied gait on a compliant quadrupedal robot. In: 2015 IEEE/RSJ International Conference on Intelligent Robots and Systems (IROS), pp 4486–4491. IEEE
42. Deitke M, Batra D, Bisk Y, Campari T, Chang AX, Chaplot DS, Chen C, D'Arpino CP, Ehsani K, Farhadi A, et al. (2022) Retrospectives on the embodied ai workshop. arXiv preprint arXiv:2210.06849

43. Deng Y, Guo D, Guo X, Zhang N, Liu H, Sun F (2020) Mqa: Answering the question via robotic manipulation. arXiv preprint arXiv:2003.04641
44. Denzler J, Brown CM (2002) Information theoretic sensor data selection for active object recognition and state estimation. IEEE Trans Pattern Anal Mach Intell 24(2):145–157
45. Di Lallo A, Catalano M, Garabini M, Grioli G, Gabiccini M, Bicchi A (2019) Dynamic morphological computation through damping design of soft material robots: application to under-actuated grippers. In: 2019 International Conference on Robotics and Automation (ICRA), pp 5155–5161. IEEE
46. Duan J, Yu S, Tan HL, Zhu H, Tan C (2022) A survey of embodied ai: From simulators to research tasks. IEEE Trans Emerg Top Comput Intell 6(2):230–244
47. Eder M, Hisch F, Hauser H (2018) Morphological computation-based control of a modular, pneumatically driven, soft robotic arm. Adv Robot 32(7):375–385
48. Eysenbach B, Salakhutdinov RR, Levine S (2021) Robust predictable control. Adv Neural Inf Proc Syst 34:27813–27825
49. Feng D, Haase-Schütz, C., Rosenbaum L, Hertlein H, Glaeser C, Timm F, Wiesbeck W, Dietmayer K (2020) Deep multi-modal object detection and semantic segmentation for autonomous driving: Datasets, methods, and challenges. IEEE Trans Intell Transp Syst 22(3):1341–1360
50. Floreano D, Lipson H (2021) From individual robots to robot societies
51. Folkertsma GA, van der Schaft AJ, Stramigioli S (2015) Morphological computation in a fast-running quadruped with elastic spine. IFAC-PapersOnLine 48(13):170–175
52. Francis J, Kitamura N, Labelle F, Lu X, Navarro I, Oh J (2022) Core challenges in embodied vision-language planning. J Artif Intell Res 74:459–515
53. Füchslin RM, Dzyakanchuk A, Flumini D, Hauser H, Hunt KJ, Luchsinger RH, Reller B, Scheidegger S, Walker R (2013) Morphological computation and morphological control: steps toward a formal theory and applications. Artif Life 19(1):9–34
54. Gan C, Zhang Y, Wu J, Gong B, Tenenbaum JB (2020) Look, listen, and act: Towards audio-visual embodied navigation. In: 2020 IEEE International Conference on Robotics and Automation (ICRA), pp 9701–9707. IEEE
55. Geiger A, Lenz P, Stiller C, Urtasun R (2013) Vision meets robotics: The kitti dataset. Int J Robot Res 32(11):1231–1237
56. Gervet T, Chintala S, Batra D, Malik J, Chaplot DS (2023) Navigating to objects in the real world. Sci Robot 8(79):eadf6991
57. Ghazi-Zahedi K (2019) Morphological intelligence. Springer, Cham
58. Ghazi-Zahedi K, Haeufle DF, Montúfar G, Schmitt S, Ay N (2016) Evaluating morphological computation in muscle and dc-motor driven models of hopping movements. Front Robot AI 3:42
59. Gordon D, Kembhavi A, Rastegari M, Redmon J, Fox D, Farhadi A (2018) Iqa: Visual question answering in interactive environments. In: Proceedings of the IEEE Conference on Computer Vision and Pattern Recognition, pp 4089–4098
60. Guo H, Wu F, Qin Y, Li R, Li K, Li K (2023) Recent trends in task and motion planning for robotics: A survey. ACM Comput Surv 55(13s):1–36
61. Gupta A, Fan L, Ganguli S, Fei-Fei L (2022) Metamorph: Learning universal controllers with transformers. arXiv preprint arXiv:2203.11931
62. Gupta A, Savarese S, Ganguli S, Fei-Fei L (2021) Embodied intelligence via learning and evolution. Nat Commun 12(1):5721
63. Gupta JK, Egorov M, Kochenderfer M (2017) Cooperative multi-agent control using deep reinforcement learning. In: Autonomous Agents and Multiagent Systems: AAMAS 2017 Workshops, Best Papers, São Paulo, Brazil, May 8–12, 2017, Revised Selected Papers 16, pp 66–83. Springer, New York
64. Ha D (2019) Reinforcement learning for improving agent design. Artif Life 25(4):352–365
65. Haarnoja T, Zhou A, Abbeel P, Levine S (2018) Soft actor-critic: Off-policy maximum entropy deep reinforcement learning with a stochastic actor. In: International Conference on Machine Learning, pp 1861–1870. PMLR

66. Han X, Liu H, Sun F, Zhang X (2019) Active object detection with multistep action prediction using deep q-network. IEEE Trans Indus Inf 15(6):3723–3731
67. Hauser H, Ijspeert AJ, Füchslin RM, Pfeifer R, Maass W (2011) Towards a theoretical foundation for morphological computation with compliant bodies. Biol Cybern 105:355–370
68. Hejna D, Pinto L, Abbeel P (2020) Hierarchically decoupled imitation for morphological transfer. In: International Conference on Machine Learning, pp 4159–4171. PMLR
69. Held R, Hein A (1963) Movement-produced stimulation in the development of visually guided behavior. J Comparat Physiol Psychol 56(5):872
70. Hiller JD, Lipson H (2009) Multi material topological optimization of structures and mechanisms. In: Proceedings of the 11th Annual Conference on Genetic and Evolutionary Computation, pp 1521–1528
71. Ho VA, Liu H, Wang L, Iida F, Hirai S (2018) Special issue on 'morphological computation in soft robotics'
72. Hong S, Yoon D, Kim KE (2021) Structure-aware transformer policy for inhomogeneous multi-task reinforcement learning. In: International Conference on Learning Representations
73. Horii Y, Inoue K, Nishikawa S, Nakajima K, Niiyama R, Kuniyoshi Y (2021) Physical reservoir computing in a soft swimming robot. In: Artificial Life Conference Proceedings 33, vol 2021, p 92. MIT Press, Cambridge, MA 02142-1209
74. Hornby GS, Lipson H, Pollack JB (2003) Generative representations for the automated design of modular physical robots. IEEE Trans Robot Autom 19(4):703–719
75. Hornby GS, Pollack JB (2001) Body-brain co-evolution using l-systems as a generative encoding. In: Proceedings of the 3rd Annual Conference on Genetic and Evolutionary Computation, pp 868–875
76. Huang K, Guo D, Zhang X, Ji X, Liu H (2024) Stimulate the potential of robots via competition. arXiv preprint arXiv:2403.10487
77. Huang W, Mordatch I, Pathak D (2020) One policy to control them all: Shared modular policies for agent-agnostic control. In: International Conference on Machine Learning, pp 4455–4464. PMLR
78. Iida F, Giardina F (2023) On the timescales of embodied intelligence for autonomous adaptive systems. Annu Rev Control Robot Auton Syst 6(1):95–122
79. Iqbal S, Sha F (2019) Actor-attention-critic for multi-agent reinforcement learning. In: International Conference on machine learning, pp 2961–2970. PMLR
80. Jaeger H, Haas H (2004) Harnessing nonlinearity: Predicting chaotic systems and saving energy in wireless communication. Science 304(5667):78–80
81. Jain U, Weihs L, Kolve E, Farhadi A, Lazebnik S, Kembhavi A, Schwing A (2020) A cordial sync: Going beyond marginal policies for multi-agent embodied tasks. In: Computer Vision–ECCV 2020: 16th European Conference, Glasgow, UK, August 23–28, 2020, Proceedings, Part V 16, pp 471–490. Springer, New York
82. Jain U, Weihs L, Kolve E, Rastegari M, Lazebnik S, Farhadi A, Schwing AG, Kembhavi A (2019) Two body problem: Collaborative visual task completion. In: Proceedings of the IEEE/CVF Conference on Computer Vision and Pattern Recognition, pp 6689–6699
83. Jayaraman D, Grauman K (2018) End-to-end policy learning for active visual categorization. IEEE Trans Pattern Anal Mach Intell 41(7);1601–1614
84. Jiang Y, Gupta A, Zhang Z, Wang G, Dou Y, Chen Y, Fei-Fei L, Anandkumar A, Zhu Y, Fan L (2022) Vima: General robot manipulation with multimodal prompts. arXiv preprint arXiv:2210.03094
85. Jing L, Tian Y (2020) Self-supervised visual feature learning with deep neural networks: A survey. IEEE Trans Pattern Anal Mach Intell 43(11):4037–4058
86. Jing Y, Kong T (2023) Learning to explore informative trajectories and samples for embodied perception. arXiv preprint arXiv:2303.10936
87. Joachimczak M, Suzuki R, Arita T (2016) Artificial metamorphosis: Evolutionary design of transforming, soft-bodied robots. Artif Life 22(3):271–298
88. Jørgensen J (2023) Towards a soft science of soft robots. a call for a place for aesthetics in soft robotics research. ACM Trans Human Robot Interact 12(2):1–11

89. Judd E, Soter G, Rossiter J, Hauscr H (2019) Sensing through the body-non-contact object localisation using morphological computation. In: 2019 2nd IEEE International Conference on Soft Robotics (RoboSoft), pp 558–563. IEEE
90. Kapelyukh I, Vosylius V, Johns E (2023) Dall-e-bot: Introducing web-scale diffusion models to robotics. IEEE Robot Autom Lett. https://doi.org/10.1109/LRA.2023.3272516
91. Katiyar SA, Lee LY, Iida F, Nurzaman SG (2023) Energy harvesting for robots with adaptive morphology. Soft Robot 10(2):365–379
92. Kim D, Moon S, Hostallero D, Kang WJ, Lee T, Son K, Yi Y (2019) Learning to schedule communication in multi-agent reinforcement learning. arXiv preprint arXiv:1902.01554
93. Kim H, Choi J, Kim KK, Won P, Hong S, Ko SH (2021) Biomimetic chameleon soft robot with artificial crypsis and disruptive coloration skin. Nat Commun 12(1):4658
94. Koditschek DE (2021) What is robotics? why do we need it and how can we get it? Annu Rev Control Robot Auton Syst 4(1):1–33
95. Koos S, Mouret JB, Doncieux S (2012) The transferability approach: Crossing the reality gap in evolutionary robotics. IEEE Trans Evol Comput 17(1):122–145
96. Kriegman S, Blackiston D, Levin M, Bongard J (2020) A scalable pipeline for designing reconfigurable organisms. Proc Natl Acad Sci 117(4):1853–1859
97. Kriegman S, Nasab AM, Shah D, Steele H, Branin G, Levin M, Bongard J, Kramer-Bottiglio R (2020) Scalable sim-to-real transfer of soft robot designs. In: 2020 3rd IEEE International Conference on Soft Robotics (RoboSoft), pp 359–366. IEEE
98. Kunold L, Bock N, Rosenthal-von der Pütten A (2023) Not all robots are evaluated equally: The impact of morphological features on robots' assessment through capability attributions. ACM Trans Human Robot Interact 12(1):1–31
99. Kurenkov A, Lingelbach M, Agarwal T, Jin E, Li C, Zhang R, Fei-Fei L, Wu J, Savarese S, Martın-Martın R (2023) Modeling dynamic environments with scene graph memory. In: International Conference on Machine Learning, pp 17976–17993. PMLR
100. Kurin V, Igl M, Rocktäschel T, Boehmer W, Whiteson S (2020) My body is a cage: the role of morphology in graph-based incompatible control. arXiv preprint arXiv:2010.01856
101. Lagerstedt E, Thill S (2023) Multiple roles of multimodality among interacting agents. ACM Trans Human Robot Interact 12(2):1–13
102. Li L, Ma S, Tokuda I, Asano F, Nokata M, Tian Y, Du L (2021) Generation of efficient rectilinear gait based on dynamic morphological computation and its theoretical analysis. IEEE Robot Autom Lett 6(2):841–848
103. Li X, Guo D, Liu H, Sun F (2022) Embodied semantic scene graph generation. In: Conference on Robot Learning, pp 1585–1594. PMLR
104. Li X, Kong T, Liu H (2023) How pre-trained visual-language models benefit language-conditioned robotic manipulation
105. Liang L, Bian G, Liu H (2023) Extracting dynamic navigation goal from natural language dialogue. In: IROS. IEEE
106. Lipson H, Pollack JB (2000) Automatic design and manufacture of robotic lifeforms. Nature 406(6799):974–978
107. Liu A, Huang T, Liu X, Xu Y, Ma Y, Chen X, Maybank SJ, Tao D (2020) Spatiotemporal attacks for embodied agents. In: Computer Vision–ECCV 2020: 16th European Conference, Glasgow, UK, August 23–28, 2020, Proceedings, Part XVII 16, pp 122–138. Springer, New York
108. Liu F, Sun F, Fang B, Li X, Sun S, Liu H (2023) Hybrid robotic grasping with a soft multimodal gripper and a deep multistage learning scheme. IEEE Trans Robot 39(3):2379–2399
109. Liu H, Guo D, Sun F, Yang W, Furber S, Sun T (2020) Embodied tactile perception and learning. Brain Sci Adv 6(2):132–158
110. Liu H, Sun F, Fang B, Lu S (2017) Multimodal measurements fusion for surface material categorization. IEEE Trans Instrum Measur 67(2):246–256
111. Liu H, Sun F, Zhang X (2018) Robotic material perception using active multimodal fusion. IEEE Trans Indus Electron 66(12):9878–9886

112. Liu H, Wang F, Guo D, Liu X, Zhang X, Sun F (2020) Active object discovery and localization using sound-induced attention. IEEE Trans Indus Inf 17(3):2021–2029
113. Liu X, Guo D, Liu H, Zhang X, Sun F (2023) Visual navigation subject to embodied mismatch. IEEE Trans Cognit Dev Syst 15(4):1959–1970
114. Liu X, Li X, Guo D, Tan S, Liu H, Sun F (2022) Embodied multi-agent task planning from ambiguous instruction. Proceedings of robotics: science and systems, New York, NY, USA pp 1–14
115. Liu Y, Chen W, Bai Y, Liang X, Li G, Gao W, Lin L (2024) Aligning cyber space with physical world: A comprehensive survey on embodied ai. arXiv preprint arXiv:2407.06886
116. Luck KS, Amor HB, Calandra R (2020) Data-efficient co-adaptation of morphology and behaviour with deep reinforcement learning. In: Conference on Robot Learning, pp 854–869. PMLR
117. Luck KS, Calandra R, Mistry M (2021) What robot do i need? Fast co-adaptation of morphology and control using graph neural networks. arXiv preprint arXiv:2111.02371
118. Lund HH, Hallam J, Lee WP (1997) Evolving robot morphology. In: Proceedings of 1997 IEEE International Conference on Evolutionary Computation (ICEC'97), pp 197–202. IEEE
119. Luo H, Lin G, Yao Y, Liu F, Liu Z, Tang Z (2022) Depth and video segmentation based visual attention for embodied question answering. IEEE Trans Pattern Anal Mach Intell 45:6807–6819
120. Luo W, Sun P, Zhong F, Liu W, Zhang T, Wang Y (2019) End-to-end active object tracking and its real-world deployment via reinforcement learning. IEEE Trans Pattern Anal Mach Intell 42(6):1317–1332
121. Ma Y, Song Z, Zhuang Y, Hao J, King I (2024) A survey on vision-language-action models for embodied ai. arXiv preprint arXiv:2405.14093
122. Marza P, Matignon L, Simonin O, Batra D, Wolf C, Chaplot DS (2023) Autonerf: Training implicit scene representations with autonomous agents. arXiv preprint arXiv:2304.11241
123. Mautner C, Belew RK (2000) Evolving robot morphology and control. Artif Life Robot 4:130–136
124. McGeer T (1990) Passive dynamic walking. Int J Robot Res 9(2):62–82
125. Medvet E, Bartoli A, Pigozzi F, Rochelli M (2021) Biodiversity in evolved voxel-based soft robots. In: Proceedings of the Genetic and Evolutionary Computation Conference, pp 129–137
126. Mnih V, Kavukcuoglu K, Silver D, Rusu AA, Veness J, Bellemare MG, Graves A, Riedmiller M, Fidjeland AK, Ostrovski G, et al. (2015) Human-level control through deep reinforcement learning. Nature 518(7540):529–533
127. Mondal WU, Agarwal M, Aggarwal V, Ukkusuri SV (2022) On the approximation of cooperative heterogeneous multi-agent reinforcement learning (marl) using mean field control (mfc). J Mach Learn Res 23(129):1–46
128. Muratore F, Ramos F, Turk G, Yu W, Gienger M, Peters J (2022) Robot learning from randomized simulations: A review. Front Robot AI, 31
129. Musha A, Daihara M, Shigemune H, Sawada H (2020) Morphological computation of skin focusing on fingerprint structure. In: Artificial Neural Networks and Machine Learning–ICANN 2020: 29th International Conference on Artificial Neural Networks, Bratislava, Slovakia, September 15–18, 2020, Proceedings, Part II 29, pp 470–481. Springer, New York
130. Nakajima K (2020) Physical reservoir computing—an introductory perspective. Jpn J Appl Phys 59(6):060501
131. Nakajima K, Hauser H, Kang R, Guglielmino E, Caldwell DG, Pfeifer R (2013) A soft body as a reservoir: case studies in a dynamic model of octopus-inspired soft robotic arm. Front Comput Neurosci 7:91
132. Nakajima K, Hauser H, Li T, Pfeifer R (2015) Information processing via physical soft body. Sci Rep 5(1):10487
133. Neftci EO, Averbeck BB (2019) Reinforcement learning in artificial and biological systems. Nat Mach Intell 1(3):133–143

134. Nelson AL, Barlow GJ, Doitsidis L (2009) Fitness functions in evolutionary robotics: A survey and analysis. Robot Auton Syst 57(4):345–370
135. Newbury R, Gu M, Chumbley L, Mousavian A, Eppner C, Leitner J, Bohg J, Morales A, Asfour T, Kragic D, et al. (2023) Deep learning approaches to grasp synthesis: A review. IEEE Trans Robot 39(5):3994–4015
136. Nguyen NH, Ngo TD, Nguyen DQ, et al. (2020) Contact distance estimation by a soft active whisker sensor based on morphological computation. In: 2020 8th IEEE RAS/EMBS International Conference for Biomedical Robotics and Biomechatronics (BioRob), pp 322–327. IEEE
137. Nurzaman SG, Wang L, Iida F, Lipton J, Floreano D, Rus D (2020) Design optimization of soft robots [from the guest editors]. IEEE Robot Autom Mag 27(4):10–11
138. Nygaard TF, Martin CP, Howard D, Torresen J, Glette K (2021) Environmental adaptation of robot morphology and control through real-world evolution. Evol Comput 29(4):441–461
139. Nygaard TF, Martin CP, Torresen J, Glette K (2019) Self-modifying morphology experiments with dyret: Dynamic robot for embodied testing. In: 2019 International Conference on Robotics and Automation (ICRA), pp 9446–9452. IEEE
140. Nygaard TF, Samuelsen E, Glette K (2017) Overcoming initial convergence in multi-objective evolution of robot control and morphology using a two-phase approach. In: Applications of Evolutionary Computation: 20th European Conference, EvoApplications 2017, Amsterdam, The Netherlands, April 19–21, 2017, Proceedings, Part I 20, pp 825–836. Springer, New York
141. Open X-Embodiment Collaboration (2023) Open x-embodiment: Robotic learning datasets and rt-x models
142. Pagliuca P, Nolfi S (2022) The dynamic of body and brain co-evolution. Adapt Behav 30(3):245–255
143. Paletta L, Pinz A (2000) Active object recognition by view integration and reinforcement learning. Robot Auton Syst 31(1–2):71–86
144. Pan X, Garg A, Anandkumar A, Zhu Y (2021) Emergent hand morphology and control from optimizing robust grasps of diverse objects. In: 2021 IEEE International Conference on Robotics and Automation (ICRA), pp 7540–7547. IEEE
145. Patel S, Wani S, Jain U, Schwing AG, Lazebnik S, Savva M, Chang AX (2021) Interpretation of emergent communication in heterogeneous collaborative embodied agents. In: Proceedings of the IEEE/CVF International Conference on Computer Vision, pp 15953–15963
146. Pathak D, Lu C, Darrell T, Isola P, Efros AA (2019) Learning to control self-assembling morphologies: a study of generalization via modularity. Adv Neural Inf Proc Syst 32. https://doi.org/10.48550/arXiv.1902.05546
147. Patten T, Zillich M, Fitch R, Vincze M, Sukkarieh S (2015) Viewpoint evaluation for online 3-d active object classification. IEEE Robot Autom Lett 1(1):73–81
148. Paul C (2006) Morphological computation: A basis for the analysis of morphology and control requirements. Robot Auton Syst 54(8):619–630
149. Peng H, Yang H, Li F, Yang C, Song N (2024) A unified framework for mechanical modeling and control of tensegrity robots. Mech Mach Theory 191:105498
150. Pervan A, Murphey TD (2020) Algorithmic design for embodied intelligence in synthetic cells. IEEE Trans Autom Sci Eng 18(3):864–875
151. Pfeifer R, Bongard J (2006) How the body shapes the way we think: a new view of intelligence. MIT Press, Cambridge, MA
152. Pfeifer R, Iida F, Lungarella M (2014) Cognition from the bottom up: on biological inspiration, body morphology, and soft materials. Trends Cognit Sci 18(8):404–413
153. Pfeifer R, Lungarella M, Iida F (2007) Self-organization, embodiment, and biologically inspired robotics. Science 318(5853):1088–1093
154. Pinto L, Gandhi D, Han Y, Park YL, Gupta A (2016) The curious robot: Learning visual representations via physical interactions. In: Computer Vision–ECCV 2016: 14th European Conference, Amsterdam, The Netherlands, October 11–14, 2016, Proceedings, Part II 14, pp 3–18. Springer, New York

155. Polani D (2011) An informational perspective on how the embodiment can relieve cognitive burden. In: 2011 IEEE Symposium on Artificial Life (ALIFE), pp 78–85. IEEE
156. Rosser K, Kok J, Chahl J, Bongard J (2020) Sim2real gap is non-monotonic with robot complexity for morphology-in-the-loop flapping wing design. In: 2020 IEEE International Conference on Robotics and Automation (ICRA), pp 7001–7007. IEEE
157. Roy N, Posner I, Barfoot T, Beaudoin P, Bengio Y, Bohg J, Brock O, Depatie I, Fox D, Koditschek D, et al. (2021) From machine learning to robotics: Challenges and opportunities for embodied intelligence. arXiv preprint arXiv:2110.15245
158. Rückert EA, Neumann G (2013) Stochastic optimal control methods for investigating the power of morphological computation. Artif Life 19(1):115–131
159. Russell SJ (2010) Artificial intelligence a modern approach. Pearson Education, London
160. Samuelsen E, Glette K (2015) Real-world reproduction of evolved robot morphologies: Automated categorization and evaluation. In: Applications of Evolutionary Computation: 18th European Conference, EvoApplications 2015, Copenhagen, Denmark, April 8–10, 2015, Proceedings 18, pp 771–782. Springer, New York
161. Sanchez-Gonzalez A, Heess N, Springenberg JT, Merel J, Riedmiller M, Hadsell R, Battaglia P (2018) Graph networks as learnable physics engines for inference and control. In: International Conference on Machine Learning, pp 4470–4479. PMLR
162. Sarch G, Fang Z, Harley AW, Schydlo P, Tarr MJ, Gupta S, Fragkiadaki K (2022) Tidee: Tidying up novel rooms using visuo-semantic commonsense priors. In: European Conference on Computer Vision, pp 480–496. Springer, New York
163. Savva M, Kadian A, Maksymets O, Zhao Y, Wijmans E, Jain B, Straub J, Liu J, Koltun V, Malik J, et al. (2019) Habitat: A platform for embodied ai research. In: Proceedings of the IEEE/CVF International Conference on Computer Vision, pp 9339–9347
164. Schaff C, Sedal A, Walter MR (2022) Soft robots learn to crawl: Jointly optimizing design and control with sim-to-real transfer. arXiv preprint arXiv:2202.04575
165. Schaff C, Yunis D, Chakrabarti A, Walter MR (2019) Jointly learning to construct and control agents using deep reinforcement learning. In: 2019 International Conference on Robotics and Automation (ICRA), pp 9798–9805. IEEE
166. Schmid L, Abate M, Chang Y, Carlone L (2024) Khronos: A unified approach for spatio-temporal metric-semantic slam in dynamic environments. In: Proc. of Robotics: Science and Systems
167. Schulman J, Wolski F, Dhariwal P, Radford A, Klimov O (2017) Proximal policy optimization algorithms. arXiv preprint arXiv:1707.06347
168. Shen B, Xia F, Li C, Martín-Martín R, Fan L, Wang G, Pérez-D'Arpino C, Buch S, Srivastava S, Tchapmi L, et al. (2021) igibson 1.0: A simulation environment for interactive tasks in large realistic scenes. In: 2021 IEEE/RSJ International Conference on Intelligent Robots and Systems (IROS), pp 7520–7527. IEEE
169. Shoeleh F, Asadpour M (2017) Graph based skill acquisition and transfer learning for continuous reinforcement learning domains. Pattern Recognit Lett 87:104–116
170. Shui L, Zhu L, Yang Z, Liu Y, Chen X (2017) Energy efficiency of mobile soft robots. Soft Matter 13(44):8223–8233
171. Sihite E, Kalantari A, Nemovi R, Ramezani A, Gharib M (2023) Multi-modal mobility morphobot (m4) with appendage repurposing for locomotion plasticity enhancement. Nat Commun 14(1):3323
172. Sims K (1994) Evolving 3d morphology and behavior by competition. Artif Life 1(4):353–372
173. Sitti M (2021) Physical intelligence as a new paradigm. Extreme Mech Lett 46:101340
174. Stanley KO, Miikkulainen R (2003) A taxonomy for artificial embryogeny. Artif Life 9(2):93–130
175. Sun F, Liu H, Yang C, Fang B (2020) Multimodal continual learning using online dictionary updating. IEEE Trans Cognit Dev Syst 13(1):171–178

176. Suresh S, Qi H, Wu T, Fan T, Pineda L, Lambeta M, Malik J, Kalakrishnan M, Calandra R, Kaess M, et al. (2024) Neuralfeels with neural fields: Visuotactile perception for in-hand manipulation. Sci Robot 9(96):eadl0628
177. Tan Q, Zhang X, Liu H, Jiao S, Zhou M, Li J (2021) Multimodal dynamics analysis and control for amphibious fly-drive vehicle. IEEE/ASME Trans Mechatron 26(2):621–632
178. Tan S, Ge M, Guo D, Liu H, Sun F (2023) Knowledge-based embodied question answering. IEEE Trans Pattern Anal Mach Intell. https://doi.org/10.48550/arXiv.2109.07872
179. Tan S, Guo D, Liu H, Zhang X, Sun F (2022) Embodied scene description. Autonomous Robots, 1–23
180. Tan S, Xiang W, Liu H, Guo D, Sun F (2020) Multi-agent embodied question answering in interactive environments. In: Computer Vision–ECCV 2020: 16th European Conference, Glasgow, UK, August 23–28, 2020, Proceedings, Part XIII 16, pp 663–678. Springer, New York
181. Tan S, Xiang W, Liu H, Guo D, Sun F (2020) Multi-agent embodied question answering in interactive environments. In: Computer Vision–ECCV 2020: 16th European Conference, Glasgow, UK, August 23–28, 2020, Proceedings, Part XIII 16, pp 663–678. Springer, New York
182. Tanaka K, Yang SH, Tokudome Y, Minami Y, Lu Y, Arie T, Akita S, Takei K, Nakajima K (2021) Flapping-wing dynamics as a natural detector of wind direction. Adv Intell Syst 3(2):2000174
183. Thrun S, Mitchell TM (1995) Lifelong robot learning. Robot Auton Syst 15(1–2):25–46
184. Turing AM (2009) Computing machinery and intelligence. Springer, New York
185. Vaswani A (2017) Attention is all you need. Adv Neural Inf Proc Syst. https://doi.org/10.48550/arXiv.1706.03762
186. Veiga T, Renoux J (2023) From reactive to active sensing: a survey on information gathering in decision-theoretic planning. ACM Comput Surv 55(13s):1–22
187. Vouros GA (2022) Explainable deep reinforcement learning: state of the art and challenges. ACM Comput Surv 55(5):1–39
188. Wang H, Wang W, Zhu X, Dai J, Wang L (2021) Collaborative visual navigation. arXiv preprint arXiv:2107.01151
189. Wang H, Wang Y, Zhong F, Wu M, Zhang J, Wang Y, Dong H (2023) Learning semantic-agnostic and spatial-aware representation for generalizable visual-audio navigation. IEEE Robot Autom Lett 8(6):3900–3907
190. Wang J, Liu H (2023) Self-supervised embodied learning for semantic segmentation. In: ICDL
191. Wang L, Zhang X, Xv B, Zhang J, Fu R, Wang X, Zhu L, Ren H, Lu P, Li J, et al. (2022) Interfusion: Interaction-based 4d radar and lidar fusion for 3d object detection. In: 2022 IEEE/RSJ International Conference on Intelligent Robots and Systems (IROS), pp 12247–12253. IEEE
192. Wang T, Liao R, Ba J, Fidler S (2018) Nervenet: Learning structured policy with graph neural networks. In: International Conference on Learning Representations
193. Wang T, Zhou Y, Fidler S, Ba J (2019) Neural graph evolution: Towards efficient automatic robot design. arXiv preprint arXiv:1906.05370
194. Wang W, Li X, Dong Y, Xie J, Guo D, Liu H (2023) Natural language instruction understanding for robotic manipulation: a multisensory perception approach. In: 2023 IEEE International Conference on Robotics and Automation (ICRA), pp 9800–9806. IEEE
195. Wang Y, Tian Y, Chen J, Xu K, Ding X (2024) A survey of visual slam in dynamic environment: the evolution from geometric to semantic approaches. IEEE Trans Instrument Measur 73:Article ID 2523221
196. Watkins CJCH (1989) Learning from delayed rewards. Ph.D Thesis
197. Weihs L, Deitke M, Kembhavi A, Mottaghi R (2021) Visual room rearrangement. In: Proceedings of the IEEE/CVF Conference on Computer Vision and Pattern Recognition, pp 5922–5931

198. Weihs L, Kembhavi A, Ehsani K, Pratt SM, Han W, Herrasti A, Kolve E, Schwenk D, Mottaghi R, Farhadi A (2021) Learning generalizable visual representations via interactive gameplay. ICLR. https://doi.org/10.48550/arXiv.1912.08195
199. Whitman J, Travers M, Choset H (2023) Learning modular robot control policies. IEEE Trans Robot. https://doi.org/10.48550/arXiv.2105.10049
200. Wijmans E, Datta S, Maksymets O, Das A, Gkioxari G, Lee S, Essa I, Parikh D, Batra D (2019) Embodied question answering in photorealistic environments with point cloud perception. In: Proceedings of the IEEE/CVF Conference on Computer Vision and Pattern Recognition, pp 6659–6668
201. Woodward MA, Sitti M (2018) Morphological intelligence counters foot slipping in the desert locust and dynamic robots. Proc Natl Acad Sci 115(36):E8358–E8367
202. Wu J, Tang SY, Fang T, Li W, Li X, Zhang S (2018) A wheeled robot driven by a liquid-metal droplet. Adv Mater 30(51):1805039
203. Wu T, Zhong F, Geng Y, Wang H, Zhu Y, Wang Y, Dong H (2022) Grasparl: Dynamic grasping via adversarial reinforcement learning. arXiv preprint arXiv:2203.02119
204. Wu W, Chang T, Li X, Yin Q, Hu Y (2024) Vision-language navigation: a survey and taxonomy. Neural Comput Appl 36(7):3291–3316
205. Xiao X, Liu B, Warnell G, Stone P (2022) Motion planning and control for mobile robot navigation using machine learning: a survey. Autonomous Robots 46(5):569–597
206. Xu J, Chen T, Zlokapa L, Foshey M, Matusik W, Sueda S, Agrawal P (2021) An end-to-end differentiable framework for contact-aware robot design. arXiv preprint arXiv:2107.07501
207. Xu Z, Wu J, Zeng A, Tenenbaum JB, Song S (2019) Densephysnet: Learning dense physical object representations via multi-step dynamic interactions. arXiv preprint arXiv:1906.03853
208. Yamashita H, Wang Z, Hirai S, Shibuya K, et al. (2017) Morphological computation in tactile sensing: The role of wrinkle. In: 2017 IEEE International Conference on Robotics and Automation (ICRA), pp 4977–4984. IEEE
209. Yang S, Nachum O, Du Y, Wei J, Abbeel P, Schuurmans D (2023) Foundation models for decision making: Problems, methods, and opportunities. arXiv preprint arXiv:2303.04129
210. Yang W, Wang X, Farhadi A, Gupta A, Mottaghi R (2018) Visual semantic navigation using scene priors. arXiv preprint arXiv:1810.06543
211. Yao R, Lin G, Xia S, Zhao J, Zhou Y (2020) Video object segmentation and tracking: A survey. ACM Trans Intell Syst Technol (TIST) 11(4):1–47
212. Yu L, Chen X, Gkioxari G, Bansal M, Berg TL, Batra D (2019) Multi-target embodied question answering. In: Proceedings of the IEEE/CVF Conference on Computer Vision and Pattern Recognition, pp 6309–6318
213. Yu Y, Huang W, Sun F, Chen C, Wang Y, Liu X (2022) Sound adversarial audio-visual navigation. arXiv preprint arXiv:2202.10910
214. Yuan Y, Song Y, Luo Z, Sun W, Kitani K (2021) Transform2act: Learning a transform-and-control policy for efficient agent design. arXiv preprint arXiv:2110.03659
215. Zhang H, Du W, Shan J, Zhou Q, Du Y, Tenenbaum JB, Shu T, Gan C (2023) Building cooperative embodied agents modularly with large language models. arXiv preprint arXiv:2307.02485
216. Zhang J, Dai L, Meng F, Fan Q, Chen X, Xu K, Wang H (2023) 3d-aware object goal navigation via simultaneous exploration and identification. In: Proceedings of the IEEE/CVF Conference on Computer Vision and Pattern Recognition, pp 6672–6682
217. Zhang X, Li Z, Gong Y, Jin D, Li J, Wang L, Zhu Y, Liu H (2022) Openmpd: An open multimodal perception dataset for autonomous driving. IEEE Trans Vehi Technol 71(3):2437–2447
218. Zhang Y, Di Z, Zhou K, Xie C, Wang X (2022) Navigation as the attacker wishes? towards building byzantine-robust embodied agents under federated learning. arXiv preprint arXiv:2211.14769
219. Zhao Q, Zhang L, Wu L, Qiao H, Liu Z (2022) A real 3d embodied dataset for robotic active visual learning. IEEE Robot Autom Lett 7(3):6646–6652

220. Zheng K, Chen X, Jenkins OC, Wang X (2022) Vlmbench: A compositional benchmark for vision-and-language manipulation. Adv Neural Inf Proc Syst 35:665–678
221. Zhong F, Bi X, Zhang Y, Zhang W, Wang Y (2023) Rspt: reconstruct surroundings and predict trajectory for generalizable active object tracking. In: Proceedings of the AAAI Conference on Artificial Intelligence, vol 37, pp 3705–3714
222. Zhong F, Sun P, Luo W, Yan T, Wang Y (2019) Ad-vat+: An asymmetric dueling mechanism for learning and understanding visual active tracking. IEEE Trans Pattern Anal Mach Intell 43(5):1467–1482
223. Zhong F, Sun P, Luo W, Yan T, Wang Y (2021) Towards distraction-robust active visual tracking. In: International Conference on Machine Learning, pp 12782–12792. PMLR
224. Zhou F, Liu H, Zhao H, Liang L (2023) Long-term object search using incremental scene graph updating. Robotica 41(3):962–975
225. Zhu Y, Mottaghi R, Kolve E, Lim JJ, Gupta A, Fei-Fei L, Farhadi A (2017) Target-driven visual navigation in indoor scenes using deep reinforcement learning. In: 2017 IEEE International Conference on Robotics and Automation (ICRA), pp 3357–3364. IEEE
226. Zou Z, Chen K, Shi Z, Guo Y, Ye J (2023) Object detection in 20 years: A survey. Proc IEEE 111(3):257–276
227. Zuo Y, Qiu W, Xie L, Zhong F, Wang Y, Yuille AL (2019) Craves: Controlling robotic arm with a vision-based economic system. In: Proceedings of the IEEE/CVF Conference on Computer Vision and Pattern Recognition, pp 4214–4223

Chapter 2
Embodied Multi-agent System

Abstract In this chapter, we demonstrate the background and motivation of embodied multi-agent system. We explain the importance and challenges of embodied multi-agent collaboration. We also provide a survey of the current research progress including multi-agent strategy learning and multi-agent embodied tasks. Then we introduce the common simulation platforms that can support embodied multi-agent collaboration.

2.1 Background and Motivation

Multi-agent collaboration is very popular in the natural world. The important motivating examples have been disclosed in [27, 34], which demonstrate the bridging capability of the ant colony. In Fig. 2.1, we find a swarm of ants can help each other to bridge the gap. After a while, the gap increases, and the ant swarm encounters a short failure and then rapidly recovers. This example shows the strong capability of multi-agent collaboration, which motivates us to shift to the more general embodied multi-agent systems, especially, the multi-robot system.

In recent years, multi-robot collaboration has attracted more and more attention and is widely applied in real scenes. For example, in warehousing and logistics, industrial manufacturing, collaborative exploration, and other scenes, multiple robots cooperate to complete complex tasks that are difficult for a single robot to complete. Multi-robot collaboration requires multiple robots to plan cooperation actions to complete specified tasks with other robots, considering the states of other robots and their interaction with the environment. The key to this process is to construct the relationship between the perception information obtained from the interaction with the environment and the execution of the actions. Thus, embodied multi-agent collaboration has become a vital direction in studying the perception, action, and learning in embodied multi-agent collaboration tasks.

H. Liu et al., *Embodied Multi-Agent Systems*, Machine Learning: Foundations, Methodologies, and Applications, https://doi.org/10.1007/978-981-96-5871-8_2

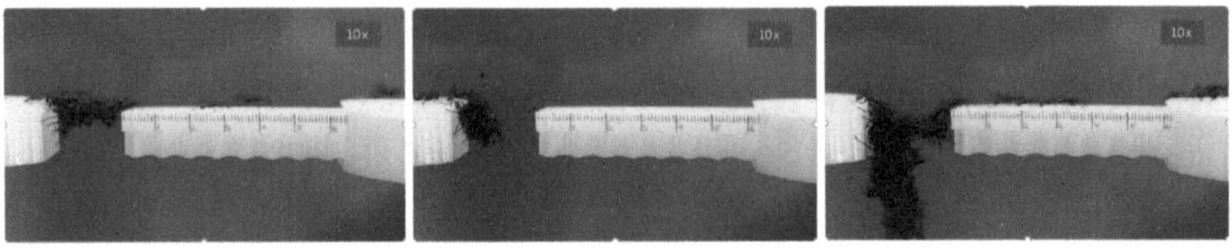

Fig. 2.1 A motivating example for bridging

2.2 Recent Progress

2.2.1 Multi-agent for Embodied Tasks

With the rapid development of embodied intelligence, researchers have begun to pay attention to embodied multi-agent collaboration strategies in complex interactive visual environment. Some work has been conducted to study the multi-agent action strategy in the confrontation tasks in 3D game scenes [1, 10]. Then, two furniture moving tasks including Furnlift [12] and Furnmove [11] are proposed to explore the multi-agent cooperative framework at the action level in an embodied visual environment. Researchers have also developed the action strategy based on 3D scene reconstruction in the newly multi-agent embodied question answering task [39]. Subsequently, researchers have considered the multi-agent embodied visual semantic navigation task in which multiple agents need to cooperate to find all target objects. The learning-based method and hierarchical decision strategy are constructed to solve the navigation task in which the specific single goal of each agent is given [40] or not given [19] respectively. On this basis, researchers have further studied the multi-agent task planning problem with ambiguous instructions as input [22]. Furthermore, the heterogeneous multi-agent collaboration framework is proposed to learn the cooperative strategy among heterogeneous agents who have different capabilities [20].

Meanwhile, the communication mechanism in embodied multi-agent collaboration has also been studied. Handshake based mechanisms including Who2com [24] and When2com [23] are proposed to calculate the communication opportunity and communication object of agents respectively, thus saving communication bandwidth (Fig. 2.2). Researchers have made an in-depth analysis of the semantic meaning of communication features in embodied multi-agent visual tasks [29].

Recently, with the surprising advancement of the large language model, it has achieved remarkable results in planning and reasoning tasks, and it has also been utilized to solve the problem of multi-agent collaboration. Four communication mechanisms based on the large language model have been proposed, and the efficiency and scalability of different communication mechanisms have been analyzed in multi-agent simple cooperative tasks [3]. In the Overcooked-AI game, ProAgent [43] uses the large language model to infer the intentions of other agents to enhance the cooperation between agents. Meanwhile, researchers have utilized the

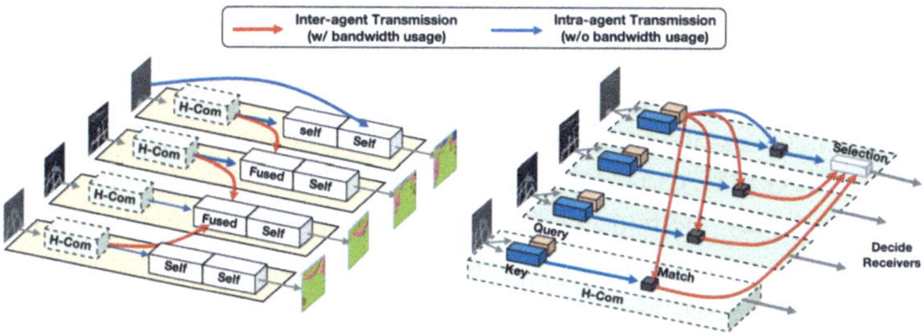

Fig. 2.2 The communication mechanism in embodied collaboration from [23]: A handshake-communication module which learns how to construct communication groups and decide when to communicate

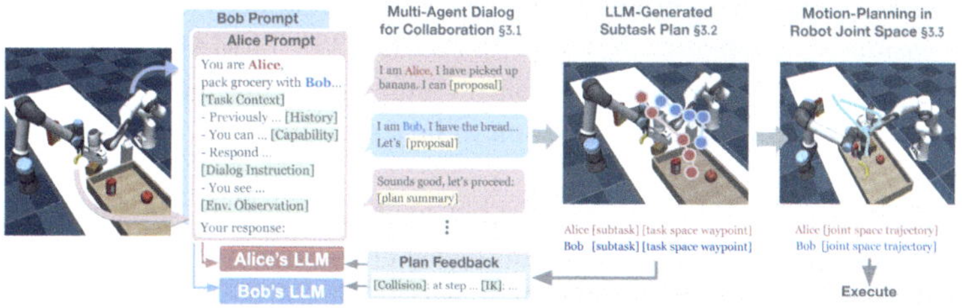

Fig. 2.3 The classical planning framework based on LLM for collaborative cooperation tasks from [26]: The framework consists of three main components: multi-agent dialog via LLM, LLM-generated sub-task planning and multi-arm motion planning

large language model to complete information sharing and collaborative behavior planning among agents in embodied environments, so that multiple agents can cooperate to complete tasks including instruction execution [14, 44], task planning [18, 42], collaborative operation [26], visual navigation [41], object rearrangement [2] and so on (Fig. 2.3). Furthermore, researchers have also constructed an ad hoc teamwork framework for more flexible heterogeneous multi-agent collaboration with the large language model that has good generalization ability [21].

2.2.2 *Multi-agent Strategy Learning*

In addition to the embodied tasks, researchers have also studied multi-agent collaboration strategy learning. Multi-agent reinforcement learning is the chief method used to solve this problem. According to the basic structure of the cooperative strategy model, these methods can be divided into two categories, namely, the value function-based methods and the Actor-Critic-based methods. The value function-based methods construct a global Q network and establish the relationship between the global Q network and the local Q network of each agent to train each agent's local Q network. These methods contain some classic structures including VDN [38], QMIX [33], QTRAN [36] and so on. The Actor-Critic-based methods train a centralized critical network, and then utilize the centralized critical network to train the actor network of each agent. These methods contain some typical methods including MADPG [25], COMA [6], MAAC [9], LICA [45] and so on. Meanwhile, the experience replay methods [7, 28] are also applied in a multi-agent setting. Opponent modeling methods [32] are developed to optimize the performance of multi-agent collaboration strategies. None of these above methods consider the communication information among multiple agents when learning cooperative strategies.

Then, researchers have started to incorporate communication information into multi-agent reinforcement learning methods. Discrete communication channels for information transmission [5] have been studied first. Subsequently, continuous channels for communication [30, 37] in multi-agent strategy learning have been further considered. Thereafter, ATOC [13], IC3Net [35], TarMAC [4] and SchedNet [15] have introduced attention mechanism into the multi-agent communication process to determine the key information to be transmitted, saving communication bandwidth.

Most multi-agent reinforcement learning methods are only suitable for simple 2D environments or game scenes. In these environments, the dimensions of scene and state features are low, so it is easy to learn the relationship between environmental observation and actions. However, these methods cannot solve the multi-agent strategy learning problem in 3D embodied visual environments. Since the perception information of agents in the 3D visual environment is visual images, the dimensions of state and scene features significantly increase. Visual images contain rich semantic and spatial information, so it is difficult to directly learn the relationship between visual perception and actions with the classic multi-agent reinforcement learning methods.

2.2.3 *Related Datasets*

Nowadays, there are several simulation platforms and datasets that can support the embodied multi-agent collaboration tasks in the simulation environment. Common datasets and platforms are introduced as follows.

AI2-THOR [16] has been proposed by Allen Institute for Artificial Intelligence, which is demonstrated in Fig. 2.4. It builds near photo-realistic 3D indoor scenes, where embodied intelligent agents can navigate and interact with objects in the scenes to complete specific embodied tasks. It contains 120 indoor single rooms, including kitchens, bedrooms, bathrooms and living rooms. It can support multi-agent tasks in both homogeneous and heterogeneous settings.

IGibson 2.0 [17] is an open-source simulation environment in the embodied AI research field developed by Stanford University, which is shown in Fig. 2.5. It supports the simulation of different household tasks and implements predicate

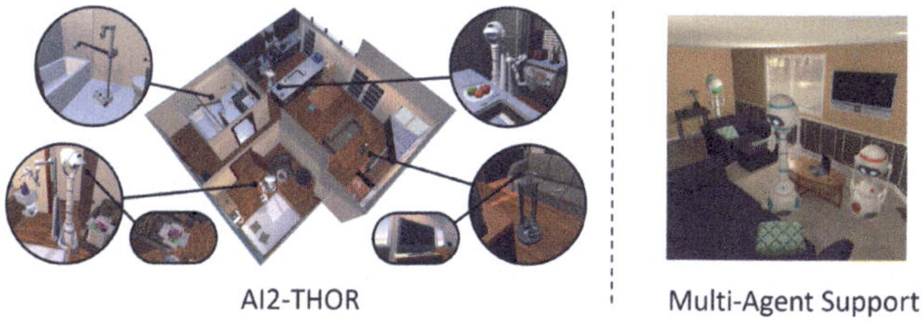

Fig. 2.4 The overview of AI2-THOR and the support for multi-agent setting. (Reproduced with permission from Ref. [16])

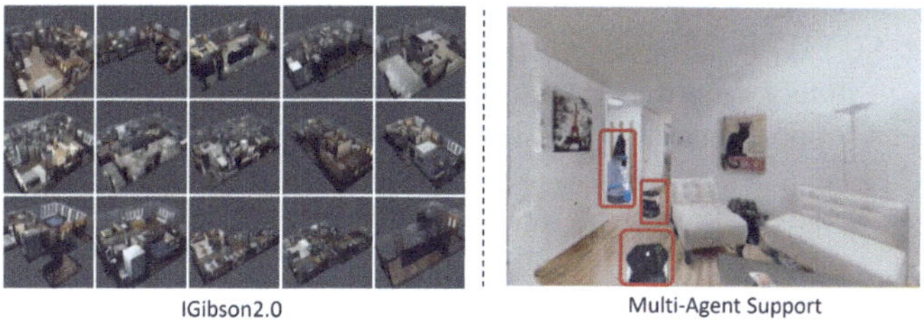

Fig. 2.5 The overview of iGibson2.0 and the support for multi-agent setting. (Reproduced with permission from Ref. [17])

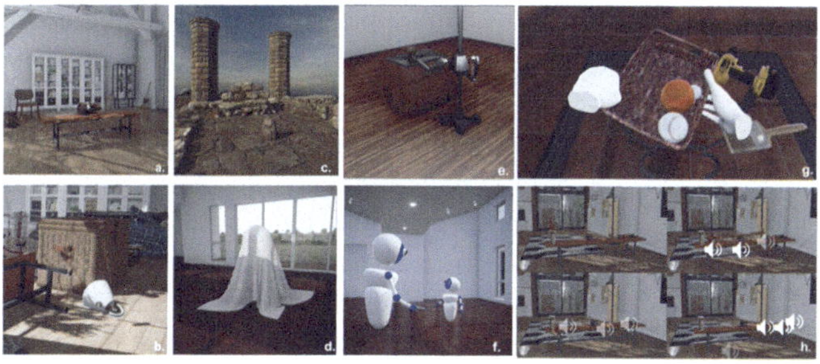

Fig. 2.6 The overview and key feature of ThreeDWorld. (Reproduced with permission from Ref. [8])

logic functions that map simulator states to logical states. It contains 15 fully interactive scenes. It also includes a virtual reality (VR) interface that provides humans the opportunity to collect demonstrations in scenes. It can support multi-agent interaction and collaboration in simulated scenes.

ThreeDWorld [8] is a platform for interactive multi-modal physical simulation built by Massachusetts Institute of Technology, which is illustrated in Fig. 2.6. It provides simulation of high-fidelity sensory data and physical interactions between mobile agents and objects in 3D environments. It can also support real-time near-photo-realistic image rendering, generative procedures for efficiently building classes of new environments, high-fidelity audio rendering, and so on. Its API enables multiple agents to interact within the simulated scenes.

Fig. 2.7 The overview of Habitat. (Reproduced with permission from Ref. [31])

Habibtat 3.0 [31] is a simulation platform supportive for collaborative human-robot tasks in indoor environments developed by Meta, which is demonstrated in Fig. 2.7. It can support accurate humanoid simulation, human-in-the-loop infrastructure, and collaborative tasks. It also provides the interface to study the human-robot collaboration in navigation and rearrangement tasks.

2.3 Architectures and Challenges

Since embodied intelligence intrinsically concerns the interaction of brain, body, and the environment, the embodied multi-agent is still a synergy of brain and body, which is illustrated in Fig. 2.8[1] It strongly depends on the cognitive emergence from the brain and behavior aggregation from the body.

Furthermore, in Fig. 2.9, we present a coarse sketch for the architecture of embodied multi-agent, which encourages the twists between morphology and behaviors, including the material-structure part and the perception-control part, respectively. All of the modules are supported by the learning component. The multi-agent structure appears at the top level, which includes the stronger and stronger brain as well as a variety of different bodies. We expect their synergy to provide effective solutions to general artificial intelligence.

Embodied multi-agent collaboration is an important research direction that combines the fields of robotics, computer vision, action decision-making, and so on. It is a challenging problem in which multiple agents need to cooperate with others to complete complex tasks in embodied environments interacting other agents and environments. Currently, the main difficulties and challenges of embodied multi-agent system are as follows.

- Effective and reasonable state feature representation and scene representation. The perception inputs of the embodied multi-agent system are always the visual

[1] This figure can be originally found in Prof. Lynne Parker's course (https://web.eecs.utk.edu/~leparker/Courses/CS594-fall08/Lectures/Oct-28-Multi-Robot-I.pdf). We make some slight changes to match the embodied scenarios.

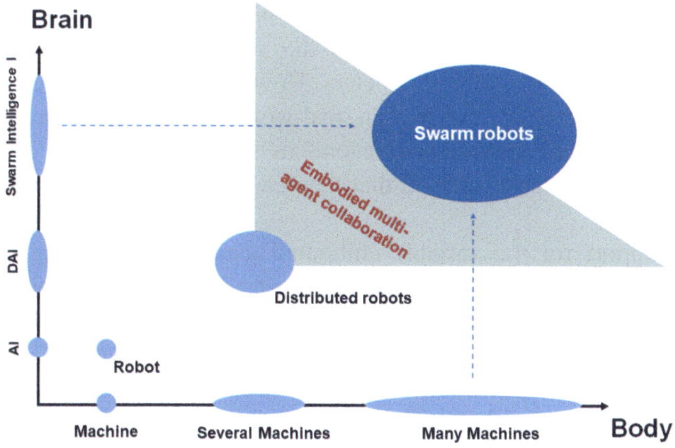

Fig. 2.8 The embodied multi-agent is a synergy of brain and body

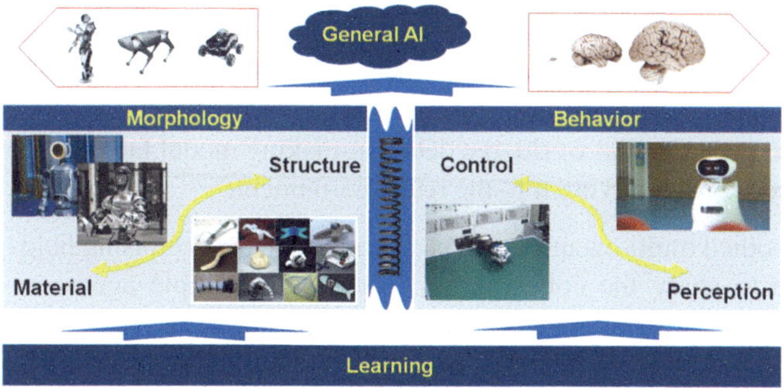

Fig. 2.9 The architecture of embodied multi-agent

observations, including visual images, depth images, etc, which contain rich semantic information and scene information with a large data dimension. In order to learn the relationship between visual perception and actions in embodied systems, it is necessary to explore robust and effective state feature representation methods and scene representation methods, so that the key information can be retained and extracted to improve the accuracy of subsequent decisions.

- Efficient collaborative decision-making framework. Embodied multi-agent collaboration requires multiple agents to perceive multi-modal information, understand scene states, and reason key links of the task being completed to generate reasonable cooperative actions. The pivot is to learn the efficient multi-agent cooperation strategy based on their current states and scene representation to improve the success rate and efficiency of the whole team to complete the task, preventing agents from generating conflicting actions or invalid actions.
- Support for the embodied heterogeneous multi-agent collaboration. The vast majority of research on embodied multi-agent system considers collaboration

strategies among homogeneous agents. In homogeneous multi-agent system, each agent has the same ability, and the strategy learning process can be simplified by sharing model parameters. However, the capabilities of different agents are likely to be different in real-world scenarios, so it is necessary to encode different capabilities of agents as well as reasonably allocate tasks according to their capabilities so that they can cooperate to complete complex tasks.

- Flexible support for the variable embodied team. Most existing research on embodied multi-agent collaboration only considers the situation in which all agents start to perform tasks at the same time. However, there may be several agents who join or quit the existing team at some timestamps after the task begins to be executed in the real-world environment. The newly joined agent does not know the historical information of the existing team and states of task completion, so it needs to infer the intention of the existing team according to the perception messages when it joins the team and then cooperates with the existing team to complete the given task.
- Reasonable strategy transfer from simulation to real-world scenarios. Most embodied multi-agent collaboration action strategies are carried out in the simulation environment. Since there always exists a large gap between the simulation and the real world, the decision-making model faces great difficulties in the process of applying it to the real environment.

The embodied multi-agent system is an important and promising field of artificial intelligence to study the collaboration strategy of multiple agents in 3D visual environments. It has not been sufficiently explored and has the above challenges. Its strong practical value and application potential in the real world make it a direction to be studied urgently. In the future, it will have broad application prospects in the fields of domestic service, warehousing and logistics, emergency rescue and, so on.

References

1. Chen B, Song S, Lipson H, Vondrick C (2020) Visual hide and seek. In: Artificial Life Conference Proceedings 32, pp 645–655. MIT Press, Cambridge, MA, USA
2. Chen J, Yu C, Zhou X, Xu T, Mu Y, Hu M, Shao W, Wang Y, Li G, Shao L (2024) Emos: Embodiment-aware heterogeneous multi-robot operating system with llm agents. arXiv preprint arXiv:2410.22662
3. Chen Y, Arkin J, Zhang Y, Roy N, Fan C (2024) Scalable multi-robot collaboration with large language models: Centralized or decentralized systems? In: 2024 IEEE International Conference on Robotics and Automation (ICRA), pp 4311–4317. IEEE
4. Das A, Gervet T, Romoff J, Batra D, Parikh D, Rabbat M, Pineau J (2019) Tarmac: Targeted multi-agent communication. In: International Conference on machine learning, pp 1538–1546. PMLR
5. Foerster J, Assael IA, De Freitas N, Whiteson S (2016) Learning to communicate with deep multi-agent reinforcement learning. Adv Neural Inf Proc Syst 29. https://doi.org/10.48550/arXiv.1911.05438

6. Foerster J, Farquhar G, Afouras T, Nardelli N, Whiteson S (2018) Counterfactual multi-agent policy gradients. In: Proceedings of the AAAI Conference on Artificial Intelligence, vol 32
7. Foerster J, Nardelli N, Farquhar G, Afouras T, Torr PH, Kohli P, Whiteson S (2017) Stabilising experience replay for deep multi-agent reinforcement learning. In: International Conference on Machine Learning, pp 1146–1155. PMLR
8. Gan C, Schwartz J, Alter S, Mrowca D, Schrimpf M, Traer J, De Freitas J, Kubilius J, Bhandwaldar A, Haber N, et al. (2020) Threedworld: A platform for interactive multi-modal physical simulation. arXiv preprint arXiv:2007.04954
9. Iqbal S, Sha F (2019) Actor-attention-critic for multi-agent reinforcement learning. In: International Conference on Machine Learning, pp 2961–2970. PMLR
10. Jaderberg M, Czarnecki WM, Dunning I, Marris L, Lever G, Castaneda AG, Beattie C, Rabinowitz NC, Morcos AS, Ruderman A, et al. (2019) Human-level performance in 3d multiplayer games with population-based reinforcement learning. Science 364(6443):859–865
11. Jain U, Weihs L, Kolve E, Farhadi A, Lazebnik S, Kembhavi A, Schwing A (2020) A cordial sync: Going beyond marginal policies for multi-agent embodied tasks. In: Computer Vision–ECCV 2020: 16th European Conference, Glasgow, UK, August 23–28, 2020, Proceedings, Part V 16, pp 471–490. Springer, New York
12. Jain U, Weihs L, Kolve E, Rastegari M, Lazebnik S, Farhadi A, Schwing AG, Kembhavi A (2019) Two body problem: Collaborative visual task completion. In: Proceedings of the IEEE/CVF Conference on Computer Vision and Pattern Recognition, pp 6689–6699
13. Jiang J, Lu Z (2018) Learning attentional communication for multi-agent cooperation. Adv Neural Inf Proc Syst 31. https://doi.org/10.48550/arXiv.1805.07733
14. Kannan SS, Venkatesh VL, Min BC (2023) Smart-llm: Smart multi-agent robot task planning using large language models. arXiv preprint arXiv:2309.10062
15. Kim D, Moon S, Hostallero D, Kang WJ, Lee T, Son K, Yi Y (2019) Learning to schedule communication in multi-agent reinforcement learning. arXiv preprint arXiv:1902.01554
16. Kolve E, Mottaghi R, Han W, VanderBilt E, Weihs L, Herrasti A, Deitke M, Ehsani K, Gordon D, Zhu Y, et al. (2017) Ai2-thor: An interactive 3d environment for visual ai. arXiv preprint arXiv:1712.05474
17. Li C, Xia F, Martín-Martín R, Lingelbach M, Srivastava S, Shen B, Vainio K, Gokmen C, Dharan G, Jain T, et al. (2021) igibson 2.0: Object-centric simulation for robot learning of everyday household tasks. arXiv preprint arXiv:2108.03272
18. Liu K, Tang Z, Wang D, Wang Z, Zhao B, Li X (2024) Coherent: Collaboration of heterogeneous multi-robot system with large language models. arXiv preprint arXiv:2409.15146
19. Liu X, Guo D, Liu H, Sun F (2022) Multi-agent embodied visual semantic navigation with scene prior knowledge. IEEE Robot Autom Lett 7(2):3154–3161
20. Liu X, Guo D, Zhang X, Liu H (2024) Heterogeneous embodied multi-agent collaboration. IEEE Robot Autom Lett 9(6):5377–5384
21. Liu X, Li P, Yang W, Guo D, Liu H (2024) Leveraging large language model for heterogeneous ad hoc teamwork collaboration. arXiv preprint arXiv:2406.12224
22. Liu X, Li X, Guo D, Tan S, Liu H, Sun F (2022) Embodied multi-agent task planning from ambiguous instruction. In: Robotics: Science and Systems
23. Liu YC, Tian J, Glaser N, Kira Z (2020) When2com: Multi-agent perception via communication graph grouping. In: Proceedings of the IEEE/CVF Conference on Computer Vision and Pattern Recognition, pp 4106–4115
24. Liu YC, Tian J, Ma CY, Glaser N, Kuo CW, Kira Z (2020) Who2com: Collaborative perception via learnable handshake communication. In: 2020 IEEE International Conference on Robotics and Automation (ICRA), pp 6876–6883. IEEE
25. Lowe R, Wu YI, Tamar A, Harb J, Pieter Abbeel O, Mordatch I (2017) Multi-agent actor-critic for mixed cooperative-competitive environments. Adv Neural Inf Proc Syst 30. https://doi.org/10.48550/arXiv.1706.02275
26. Mandi Z, Jain S, Song S (2024) Roco: Dialectic multi-robot collaboration with large language models. In: 2024 IEEE International Conference on Robotics and Automation (ICRA), pp 286–299. IEEE

27. Mlot NJ, Tovey CA, Hu DL (2011) Fire ants self-assemble into waterproof rafts to survive floods. Proc Natl Acad Sci 108(19):7669–7673
28. Omidshafiei S, Pazis J, Amato C, How JP, Vian J (2017) Deep decentralized multi-task multi-agent reinforcement learning under partial observability. In: International Conference on Machine Learning, pp 2681–2690. PMLR
29. Patel S, Wani S, Jain U, Schwing AG, Lazebnik S, Savva M, Chang AX (2021) Interpretation of emergent communication in heterogeneous collaborative embodied agents. In: Proceedings of the IEEE/CVF International Conference on Computer Vision, pp 15953–15963
30. Peng P, Wen Y, Yang Y, Yuan Q, Tang Z, Long H, Wang J (2017) Multiagent bidirectionally-coordinated nets: Emergence of human-level coordination in learning to play starcraft combat games. arXiv preprint arXiv:1703.10069
31. Puig X, Undersander E, Szot A, Cote MD, Yang TY, Partsey R, Desai R, Clegg AW, Hlavac M, Min SY, et al.(2023) Habitat 3.0: A co-habitat for humans, avatars and robots. arXiv preprint arXiv:2310.13724
32. Rabinowitz N, Perbet F, Song F, Zhang C, Eslami SA, Botvinick M (2018) Machine theory of mind. In: International conference on machine learning, pp 4218–4227. PMLR
33. Rashid T, Samvelyan M, De Witt CS, Farquhar G, Foerster J, Whiteson S (2020) Monotonic value function factorisation for deep multi-agent reinforcement learning. J Mach Learn Res 21(178):1–51
34. Reid CR, Lutz MJ, Powell S, Kao AB, Couzin ID, Garnier S (2015) Army ants dynamically adjust living bridges in response to a cost–benefit trade-off. Proc Natl Acad Sci 112(49):15113–15118
35. Singh A, Jain T, Sukhbaatar S (2018) Learning when to communicate at scale in multiagent cooperative and competitive tasks. arXiv preprint arXiv:1812.09755
36. Son K, Kim D, Kang WJ, Hostallero DE, Yi Y (2019) Qtran: Learning to factorize with transformation for cooperative multi-agent reinforcement learning. In: International Conference on Machine Learning, pp 5887–5896. PMLR
37. Sukhbaatar S, Fergus R, et al. (2016) Learning multiagent communication with backpropagation. Adv Neural Inf Proc Syst 29. https://doi.org/10.48550/arXiv.1605.07736
38. Sunehag P, Lever G, Gruslys A, Czarnecki WM, Zambaldi V, Jaderberg M, Lanctot M, Sonnerat N, Leibo JZ, Tuyls K, et al. (2017) Value-decomposition networks for cooperative multi-agent learning. arXiv preprint arXiv:1706.05296
39. Tan S, Xiang W, Liu H, Guo D, Sun F (2020) Multi-agent embodied question answering in interactive environments. In: Computer Vision–ECCV 2020: 16th European Conference, Glasgow, UK, August 23–28, 2020, Proceedings, Part XIII 16, pp 663–678. Springer, New York
40. Wang H, Wang W, Zhu X, Dai J, Wang L (2021) Collaborative visual navigation. arXiv preprint arXiv:2107.01151
41. Yu B, Kasaei H, Cao M (2023) Co-navgpt: Multi-robot cooperative visual semantic navigation using large language models. arXiv preprint arXiv:2310.07937
42. Yu W, Peng J, Ying Y, Li S, Ji J, Zhang Y (2024) Mhrc: Closed-loop decentralized multi-heterogeneous robot collaboration with large language models. arXiv preprint arXiv:2409.16030
43. Zhang C, Yang K, Hu S, Wang Z, Li G, Sun Y, Zhang C, Zhang Z, Liu A, Zhu SC, et al. (2024) Proagent: building proactive cooperative agents with large language models. In: Proceedings of the AAAI Conference on Artificial Intelligence, vol 38, pp 17591–17599
44. Zhang H, Du W, Shan J, Zhou Q, Du Y, Tenenbaum JB, Shu T, Gan C (2023) Building cooperative embodied agents modularly with large language models. arXiv preprint arXiv:2307.02485
45. Zhou M, Liu Z, Sui P, Li Y, Chung YY (2020) Learning implicit credit assignment for cooperative multi-agent reinforcement learning. Adv Neural Inf Proc Syst 33:11853–11864

Part II
Theory and Methods

This part of the book focuses on the theory and methods of collaborative active perception and interactive learning. It comprises four chapters. Chapter 3 tackles the perception-action loop in embodied multi-agent systems, and Chap. 4 solves the embodied collaboration. In Chap. 5, we shift to the multi-agent interactive learning, which includes the competitive and adversarial learning methods. Finally, in Chap. 6, we present an advanced topic: how to use large language model to enhance the collaboration performance.

Chapter 3
Perception-Action Loop in Embodied Multi-Agent System

Abstract In this chapter, we discuss the framework of the perception-action loop in the embodied multi-agent system for both the homogeneous multi-agent system and the heterogeneous multi-agent system, respectively. The basic structure of the perception-action loop includes visual perception, state feature extraction, communication, and action decision modules. The loop also has different detailed designs according to the characteristics of different tasks. We first analyze the centralized perception-action loop structure in the autonomous task assignment. We propose a dynamic, collaborative task planning framework that utilizes external knowledge sources with dynamically perceived visual information to resolve the high-level instructions and dynamically allocates the decomposed tasks to multiple agents. Next, we discuss the decentralized perception-action loop structure in collaborative object search. We develop a hierarchical decision framework based on semantic mapping, scene prior knowledge, and communication mechanism to solve this task. Then, we analyze the perception-action loop structure in the heterogeneous embodied multi-agent system in the room tidying-up task, in which heterogeneous agents collaborate with others to detect misplaced objects and place them in reasonable locations. The heterogeneous perception-action loop needs to consider the different capabilities of agents when making action decisions. We introduce a hierarchical decision model based on misplaced object detection, reasonable receptacle prediction, and handshake-based group communication mechanism for heterogeneous multi-agent room tidying-up. We also demonstrate simulation experiments of these perception-action frameworks.

3.1 Introduction

In embodied multi-agent system, multiple agents are required to perceive the scene, extract crucial information, and collaboratively execute reasonable actions to interact with the environment and complete the corresponding tasks. Therefore, the perception-action loop is the core component in the action decision and cooperative strategy of embodied multi-agent system. According to the capabilities and cooperation mechanism of agents, embodied multi-agent system can be

H. Liu et al., *Embodied Multi-Agent Systems*, Machine Learning: Foundations, Methodologies, and Applications, https://doi.org/10.1007/978-981-96-5871-8_3

divided into different categories. Specifically, based on whether each agent has the same ability and morphology, embodied multi-agent system can be classified into homogeneous systems and heterogeneous systems. In homogeneous systems, all agents have the same action capabilities and body shapes so that they can share the parameters and weights of their action decision models during strategy learning. In heterogeneous systems, agents would have different action capabilities or body shapes. When making decisions, they need to consider the difference in their abilities as well as the adaptability between the specific capability and task attributes to conduct collaboration. Based on the cooperation decision and communication mechanism, embodied multi-agent system can be divided into centralized systems and decentralized systems. In centralized systems, there exists a central node to transfer all communication messages and plan actions for all agents based on all collected information. The robustness of the centralized system is relatively poor, and if the central node is attacked, agents in the system cannot continue to cooperate with others. In decentralized systems, agents communicate and make decisions by themselves based on received information from others. Decentralized systems have better robustness, but it is more difficult for them to learn collaborative strategies than centralized systems. In this chapter, we discuss the different structures of the perception-action loop in homogeneous, heterogeneous, centralized, and decentralized settings.

The structure of this chapter is shown as follows. We demonstrate the centralized perception-action loop in homogeneous embodied multi-agent system, taking the embodied multi-agent task assignment problem as an example in Sect. 3.2. We discuss the decentralized perception-action loop in homogeneous embodied multi-agent system in the collaborative object search task in Sect. 3.3. Then, we further illustrate the perception-action loop in heterogeneous embodied multi-agent system in the collaborative room tidying-up task in Sect. 3.4.

3.2 Centralized Perception-Action Loop in Homogeneous Embodied Multi-Agent System

In this section, we analyze the centralized perception-action loop in the homogeneous embodied multi-agent system in the multi-agent task assignment problem. We introduce the embodied multi-agent task assignment problem and discuss the perception-action framework to solve this problem. Then we demonstrate the experiment settings and results in the simulation environment.

3.2.1 Problem Formulation

In the embodied multi-agent task assignment problem, multiple homogeneous agents aim to decompose the task indicated by the high-level ambiguous instruction

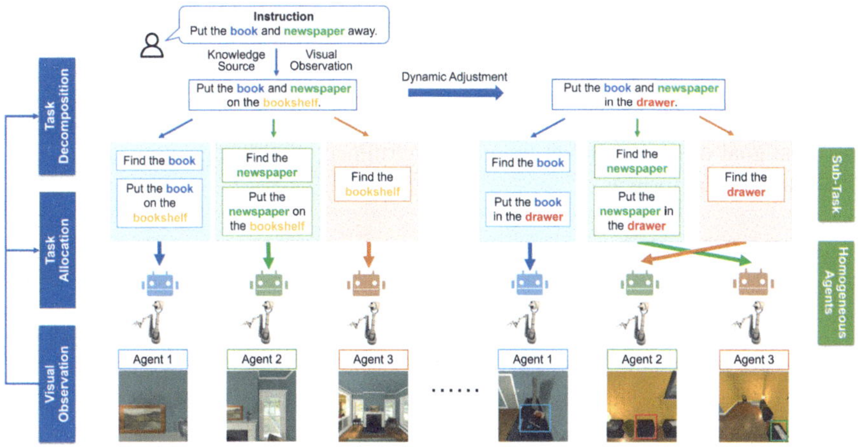

Fig. 3.1 An overview of the centralized homogeneous multi-agent task assignment problem

into several subtasks, collaboratively allocate subtasks, and cooperate to complete the entire task with egocentric visual observations, which is illustrated in Fig. 3.1.

Concretely speaking, we consider embodied agents $\{A^{(1)}, A^{(2)}, \cdots, A^{(N)}\}$, where N is the number of agents. All of the agents are homogeneous and share an action space $\mathcal{A}$, from which each agent can take an action to perform at each time instant. The state representation of the i-th agent is represented as $S_t^{(i)}$.

Given a high-level language instruction $\mathcal{L}$, it contains the main intention of the person but lacks necessary operation details to ensure the task to be performed. What we hope to address is to analyze the intention embedding in the instruction, complete the operation details using external knowledge source, and perform the task planning for practice execution. Thus, we should first transform the original instruction $\mathcal{L}$ as structural representation $\mathcal{P}$, which may contain incomplete or missing items due to the ambiguity of the instruction $\mathcal{L}$.

Therefore, at each time instant t, we have to resort to the available external knowledge source $\mathcal{K}$ and the agents' perception information to complete the necessary operation details and decompose the whole task into subtasks $\mathcal{T}_t = \{T_t^{(1)}, T_t^{(2)}, \cdots, T_t^{(K)}\}$ with temporal ordering constraints, where $T_t^{(k)}$ denotes a specific subtask, and K is the number of subtasks. This procedure can be represented as

$$\mathcal{T}_t = \mathrm{TD}(\mathcal{P}, \mathcal{S}_t),$$

where $\mathcal{S}_t = \{S_t^{(1)}, S_t^{(2)}, \cdots, S_t^{(N)}\}$ are the collected state information from the agents. Obviously, the above representations formulate a perception-aware task decomposition problem, and it results in a list of subtasks that should be performed. The next step is to allocate them to the embodied agents. The task allocation results can be represented as $\mathcal{V}_t = \{V_t^{(1)}, V_t^{(2)}, \cdots, V_t^{(N)}\}$, which can be derived as

$$\mathcal{V}_t = \mathrm{TA}(\mathcal{T}_t, \mathcal{S}_t).$$

Once the goal of each agent at time t is determined, a multi-agent search strategy can be used to perform as

$$a_t^{(i)} = \pi^{(i)}\left(S_t^{(i)}, S_t^{(i^-)}, V_t^{(i)}\right) \tag{3.1}$$

for $i = 1, 2, \cdots, N$, where $S_t^{(i^-)}$ is the state information of other agents which can be transferred via communication.

It should be noted that the goal of each agent $V_t^{(i)}$ is time-varying, and it is determined from an ambiguous instruction, which makes this problem more challenging.

3.2.2 *Perception-Action Framework*

The embodied multi-agent task assignment problem requires the agents to reason a set of subtasks from the high-level instruction and to correct and reallocate the subtasks in the exploring process dynamically. Considering these challenges, we propose a hierarchical multi-agent framework to solve the task. Our framework consists of four essential modules: scene encoder module, task planning module, which contains the task decomposition and task allocation part, action module, and communication module, as is shown in Fig. 3.2.

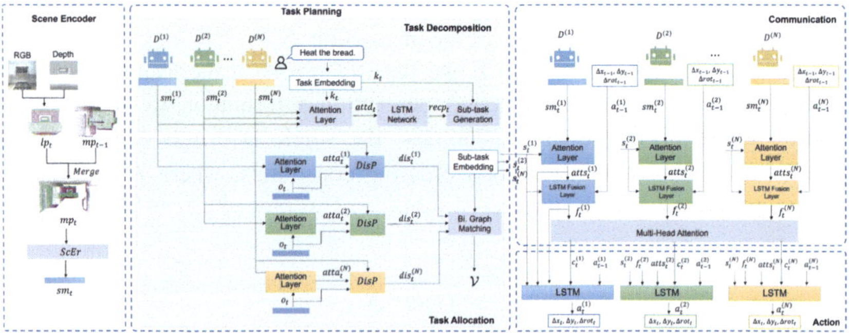

Fig. 3.2 An overview of the hierarchical perception-action framework for the embodied multi-agent task assignment problem

3.2.2.1 Scene Encoder

The scene encoder module first takes the RGB and depth observations as input to obtain the semantic map, and a pretrained method [18] is utilized to generate the semantic map feature vector. The *Swin Transformer* is used to obtain the semantic point cloud from the agent's current RGB and depth observations. Then we voxelize the semantic point cloud with a voxel size of 0.125 m × 0.125 m, and a semantic map of $C \times M \times M$ is obtained, where C indicates the number of object types, and M denotes the length of the semantic map. We maintain the egocentric semantic map that is within four meters around the agent, so the size of i-th agent's semantic map $mp_t^{(i)}$ is $C \times 64 \times 64$ at step t. Specifically, for step t, the i-th agent's semantic map $mp_t^{(i)}$ is generated by merging its current local semantic map $lp_t^{(i)}$ and the semantic map $mp_{t-1}^{(i)}$ from previous step.

Then, a scene encoder $ScEr$ is trained with a pretraining method to generate the feature vector $sm_t^{(i)}$ from the semantic map $mp_t^{(i)}$. The feature vector is supposed to retain the region information in the semantic map. The scene encoder is composed of a series of convolution layers. To train the scene encoder, we give a query to see if a specific type of object exists in a subregion of the map. The scene encoder is trained to answer this query correctly to get the reasonable semantic map feature vector. With the pretrained scene encoder $ScEr$, we could have the semantic map feature vectors as $sm_t^{(i)} = ScEr(Merge(mp_{t-1}^{(i)}, lp_t^{(i)}))$, which are the input for the task planning and action modules.

3.2.2.2 Task Planning

The task planning module is composed of the task decomposition part and the task allocation part. The task decomposition part takes the high-level instruction and the semantic map feature vectors as input for receptacle reasoning and subtask sequences generation. The task allocation part allocates the subtask sequence to each agent conditioned on its semantic map information.

Task Decomposition Given a high-level instruction, the task decomposition part first extracts the key actions and objects in the instruction, and the implicit receptacles in the instruction are predicted based on the semantic map feature vectors. At step t, we concatenate the encoding of the action and objects from the instruction to generate the task embedding k_t. And then the task embedding k_t and the current semantic map feature vectors $sm_t^{(i)}, i = 1, 2, \cdots, N$ from all agents are fed into an attention layer, and the attention $attd_t$ is obtained. We use the LSTM network to predict the implicit receptacle as $recp_t = F_d(LSTM(k_t, attd_t))$, where F_d denotes the receptacle prediction network. The given incomplete instruction can be completed with the receptacle $recp_t$. Then, the decomposed subtasks and their corresponding temporal orders are generated based on the type of the given instruction.

Task Allocation The task allocation part predicts the distance from the agents' current location to target objects in the scene, given its egocentric semantic map feature vector $sm_t^{(i)}$. At step t, given the embedding of the target objects o_t, we firstly calculate the attention $atta_t^{(i)}$ of o_t on semantic map vector $sm_t^{(i)}$. Then the distance $dis_t^{(i)}$ from the agent's current location to target objects is predicted as follows:

$$
\begin{aligned}
dis_t^{(i)} &= DisP\left(sm_t^{(i)}, LSTM\left(atta_t^{(i)}\right), o_t\right) \\
dis_t^{(i)} &= \left\{dis_t^{(i)}(1), dis_t^{(i)}(2), \cdots, dis_t^{(i)}(H)\right\},
\end{aligned}
\tag{3.2}
$$

where H is the number of target objects, and $DisP$ denotes the distance predictor which consists of linear layers. After the distances are predicted, the task allocation part utilizes a method similar to the bipartite graph matching algorithm to allocate each subtask to each agent based on their predicted distances so that the total length executed when multiple agents complete the task is the smallest. And the result of task allocation is denoted as $\mathcal{V} = F_a(dis_t^{(1)}, dis_t^{(2)}, \cdots, dis_t^{(N)})$, where F_a denotes the task allocation model. The task allocation method requires that each subtask is assigned to at least one agent for execution.

It is noted that the task decomposition and task allocation results can be adjusted dynamically. After each step t, new visual observations can be obtained after the agents execute actions based on their assigned subtasks. Then, with their updated semantic map feature vectors, the task planning module will also be updated correspondingly.

3.2.2.3 Communication

Having the subtasks allocated, the communication module exchanges the information among agents. The agent can utilize the message from other agents to facilitate itself for the following interaction module.

For each agent $A^{(i)}$, we utilize the triple $(action, object, receptacle)$ to represent the corresponding subtask for each agent, and "*blank*" is used to denote a vacant item in the subtask, such as $(find, bread, blank)$ for the subtask "Find the bread." Then we encode the triple to get the subtask embedding $s_t^{(i)}$ assigned to agent A^i at step t. The subtask $s_t^{(i)}$ and the agent's semantic map feature vector $sm_t^{(i)}$ are firstly fed into an attention layer, and the attention $atts_t^{(i)}$ is obtained. The action information of the agent is represented by the subgoal. And the subgoal of the agent in the previous step $t-1$ includes the movement distance in its egocentric x and y axis $\Delta x_{t-1}^{(i)}$ $\Delta y_{t-1}^{(i)}$, and the rotation angle $\Delta rot_{t-1}^{(i)}$. We encode the previous subgoal to get the action embedding $a_{t-1}^{(i)}$. And then the attention $atts_t^{(i)}$, the corresponding subtask $s_t^{(i)}$, and the subgoal of the agent in the previous step $a_{t-i}^{(i)}$ are fed into the LSTM fusion layer, and the fused feature $f_t^{(i)}$ is obtained.

If we view the multi-agents as a sequence of agents, then the communication between each pair of agents is similar to the self-attention mechanism, which is widely used in transformer-based methods. Therefore, we use the multi-head attention layer for the information communication, which takes the fused feature $f_t^{(i)}$ from multiple agents as input and calculates the attention of each agent's state information on other agents' states separately to generate the communication embedding as

$$\left(c_t^{(1)}, c_t^{(2)} \cdots, c_t^{(N)}\right) = F_{mha}\left(f_t^{(i)}, f_t^{(2)} \cdots, f_t^{(N)}\right),$$

where F_{mha} denotes the multi-head attention network. Each agent can selectively utilize the beneficial information from others for the following action module through the attention mechanism. Meanwhile, we modify the traditional multi-head attention structure with the Softmax activation function to make the communication module more convenient to extend to the multi-agent system with a larger number of agents.

3.2.2.4 Action Module

The action module leverages the communication information, semantic information, subtask information, and previous subgoal to predict the next subgoal for the agent. After obtaining the subgoal, the low-level action generation policy is employed to generate the navigation actions for the agent to execute in the environment.

We train a subgoal predictor F_p to generate the next subgoal $a_t^{(i)}$, which includes the egocentric movement distance in the x and y axis, namely $\Delta x_t^{(i)}$ and $\Delta y_t^{(i)}$, the rotation angle $\Delta rot_t^{(i)}$, and the probability of "Stop" action $stop_t^{(i)}$. Given $f_t^{(i)}$, $s_t^{(i)}$, $a_{t-1}^{(i)}$, $atts_t^{(i)}$, and c_t^i, the next subgoal $a_t^{(i)} = F_p(LSTM(f_t^{(i)}, s_t^{(i)}, a_{t-1}^{(i)}, atts_t^{(i)}, c_t^{(i)})$ is generated. If the agent performs "stop" action, the model judges whether the target object is found. If it is found, a specific manipulation action such as "Pick" is performed, and the current subtask is switched to the next one. If the current subtask is the last one, then the entire task is completed.

To navigate to the subgoal, in the simulation environments, a low-level action strategy based on the shortest path algorithm is utilized to make agents navigate from the current subgoal to the next subgoal. We can then apply this structure to the real-world scenarios. In real-world experiments, agents can navigate from the current location to the predicted subgoal based on their own localization, obstacle avoidance, and path planning system. In this way, we can bridge the difference between the simulation and the real world to the most.

3.2.2.5 Training Methodology

Since both the task planning module and the action module take the semantic map feature vector as input and the training process of the two modules is relatively independent, we train these two modules separately. We utilize the supervised learning algorithm to train the task planning module and the imitation learning algorithm to train the action model.

The task decomposition part in the task planning module predicts the implicit receptacles, and the task allocation part predicts the distance between the agent and the target objects in the scene. The ground truth of the implicit receptacles can be obtained from the true task triple. The true distance between the agent's location and all objects can be obtained from the AI2-THOR simulator. During the training process, the task allocation part is trained to predict the distance from the agent's location to all objects. During the inference process, the agents only predict the distance to target objects. The loss function of the task planning module is defined as follows:

$$\begin{aligned} Loss_{task} &= \alpha Loss_{recep} + \beta Loss_{dis}, \\ Loss_{dis} &= Loss_{seen} + \lambda Loss_{unseen}, \end{aligned} \tag{3.3}$$

where $Loss_{recep}$ denotes the entropy loss between the predicted receptacles and the true receptacles, and $Loss_{dis}$ denotes the sum of MSE loss of distance to objects. $Loss_{dis}$ contains the regression loss of the predicted distance to seen objects $Loss_{seen}$ and the regression loss of the distance to unseen objects $Loss_{unseen}$. α, β, and λ are the hyper-parameters to control the training process.

The action module task predicts the next subgoal specified by the grid of 0.25 meters given the specific subtask embedding and semantic map feature vectors. The subgoal is at most 1 meter from the original position on the egocentric x and y axes, that is, four grids in the simulator. Meanwhile, each subgoal also contains the rotation angle, which is specified in units of 90°, that is, the rotation in the new subgoal can take the values of 0, 90, 180, 270. The subgoal navigation demonstrations are generated by the heuristic shortest path algorithm. For each subtask, we find the viewpoints in which the agent can interact with the target object and select several viewpoints with the largest bounding box area in the RGB observations as the target location. Then we generate the shortest path in the subgoal level. Since the movement in egocentric x and y axes and rotation degrees are relatively independent during navigation, we predict the number of moving grids in the two directions and the rotation angle, respectively. We also predict the probability of the action "Stop" to monitor the progress of the subtask. The loss function of the action module is defined as follows:

$$\begin{aligned} Loss_{sub-goal} = \gamma Loss_{x_{loc}} &+ \delta Loss_{y_{loc}} \\ &+ \theta Loss_{rot} + \phi Loss_{stop}, \end{aligned} \tag{3.4}$$

where $Loss_{x_{loc}}$ denotes the cross entropy loss between the number of predicted subgoal movement grids in egocentric x axis direction, and that in demonstrations, $Loss_{y_{loc}}$ denotes the cross entropy loss between the number of predicted subgoal movement grids in egocentric y axis, and that in demonstrations, $Loss_{rot}$ indicates the cross entropy loss between the predicted rotation angle, and that in demonstrations, $Loss_{stop}$ denotes the binary of cross entropy loss of the binary classification problem whether the agent stops or not. γ, δ, θ, and ϕ are the hyper-parameters to control the training process.

3.2.3 *Simulation Experiments*

3.2.3.1 Dataset

To evaluate the proposed perception-action loop in embodied task assignment task, we build a specific scenario benchmark simulation dataset to evaluate the proposed framework. In this problem, we restrict ourselves to deal with the ambiguous instructions from which we may infer the receptacles for some objects.

To develop a benchmark, we resort to ALFRED [17] dataset in AI2-THOR simulator [10]. ALFRED is a recently developed benchmark for interpreting grounded instructions for everyday tasks. Based on the ALFRED dataset, reasonable and abstract high-level instructions in which the target object is implicit and needs to be inferred can be generated. However, ALFRED does not provide ambiguous instructions, and it cannot be directly used for multi-agent exploration.

In this chapter, we use the instructions in ALFRED to extract the necessary information and construct a new benchmark with AI2-THOR, which supports multi-agent collaboration. Concretely speaking, we extract five fundamental types of behaviors *heat*, *cool*, *clean*, *put*, and *throw* in the high-level instructions. In addition, we extract the relationships between object properties and behaviors in ALFRED, such as *bread can be heated* and *book cannot be heated or cleaned*. We also preserve the placement constraints in ALFRED between small objects and receptacles, which imposes the constraints on whether a certain type of object could be placed on or inside another type of receptacle object. For example, the reasonable receptacles of *book* can be *bookshelf*, *desk*, and *countertop* rather than *garbage can*. This provides solutions to design reasonable ambiguous instructions.

Then, we use the AI2-THOR environment to design our tasks. The constructed dataset consists of three types of multi-agent tasks indicated by high-level instructions, in which the target objects and operation information may be implicit and need to be reasoned during the process of task execution. As illustrated in Fig.3.3, the three types of tasks are generated based on the fundamental actions, object properties, and placement constraints. When generating high-level instructions for each type of tasks, we select each type of action, preserve objects meeting object properties constraints for the specific action, and obtain the most likely reasonable receptacles for the specific pairs of action and object to form the triple

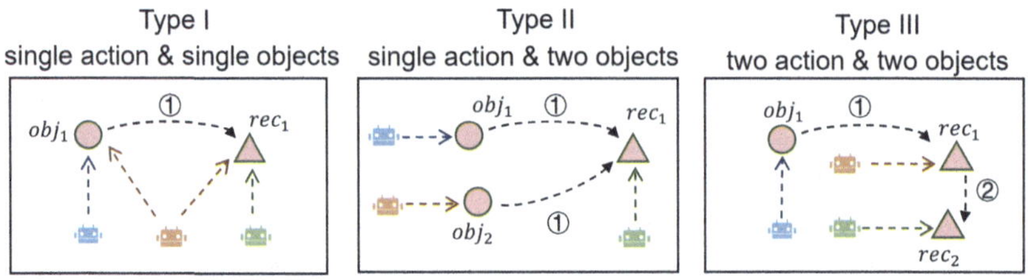

Fig. 3.3 Three types of instructions in our dataset

$(action, object, receptacle)$. In particular, we sort the reasonable receptacles in descending order according to the frequency of the triples with the specific action and object pair that appeared in ALFRED instructions. For the fixed action and object, the candidate triples are utilized to generate the high-level instructions.

For the Type I task, the high-level instructions contain one action and one object. Instructions are generated directly based on obtained triplets as well as several language rules, and implicit receptacles do not appear explicitly in instructions. For instance, in the instruction *"Heat the bread,"* the receptacle *microwave* needs to be reasoned.

For the Type II task, the high-level instructions include one action and two objects that can be executed by that action. We select two objects with a common receptacle for a specific action from candidate triples and generate high-level instructions in a similar way, such as *"Heat the tomato and the bread."*

For the Type III task, the high-level instructions contain two actions and two objects, in which two actions have a temporal order, and the object processed by the first action is also the target object of the second action *put*. After the first action is completed, the operated object needs to be placed on another receptacle, so we filter to get receptacles of the second action according to the placement constraints. The instructions are generated using "and" to connect two action instructions, and the receptacle of the first action is implicit. For instance, in the instruction *"Heat the bread and put it on the countertop,"* the *bread* needs to be put on the *counterTop* after being heated in the *microwave*.

It is noted that the receptacles are implicit in all three types of tasks, and they need to be inferred when the task is processed. Meanwhile, all three types of tasks can be decomposed into subtasks that are executed by multiple agents.

To facilitate evaluating the dynamic task allocation strategy, we utilize the object mask to perform data argumentation. For the sorted receptacles, we mask out the top three objects in the scene respectively and select another object of the highest frequency to form new triples and generate new tasks of the three types. The correct inferred receptacles are different from the task without masked objects, but the generated high-level instructions are the same since the receptacles are implicit in instructions. When the agent cannot see the top 1 masked receptacle during the task process, it needs to dynamically adjust the inferred results to find other possible receptacles. For example, in the instruction *"Put the book away,"* the most likely receptacle *Drawer* is masked, and the agent needs to re-reason the receptacle to be *Cabinet* after finding no *Drawer* in the scene. Meanwhile, it is noted that to make the generated dataset more natural and flexible, we extend the original dataset to a larger one, in which we utilize natural language processing methods to generate richer daily activity instructions through synonym generation, syntax analysis, and sentence pattern conversion.

The expert demonstrations are also generated, which consist of the task decomposition results, subtask allocation results, and actions to be taken at each step for every agent. The task decomposition results are generated with specific language rules and the ground truth task triples. We consider the situation where three agents are in the environment, and the locations of the three agents are initialized under the constraints that the distance between multiple agents is greater than 1m and the distance between agents and the same target object is also greater than 1m, ensuring allocation results reasonable. The allocation results are generated with the task decomposition results and agents' initialization locations. The navigation actions are generated by the shortest path algorithm at the subgoal level with the rotation information.

From the above description, we can see that though the developed benchmark is dependent on AI2-THOR, it is significantly different from ALFRED. Finally, we divide the full dataset, including 120 scenes, into three nonoverlapping splits, namely Train, Validation, and Test. The Train split contains 80 scenes, while the Validation and Test splits contain 20 unseen scenes, respectively. The corresponding information is shown in Table 3.1.

3.2.3.2 Experiment Settings

To comprehensively evaluate the proposed method, we design two settings *Easy* and *Hard* using the developed dataset. The main difference lies in the layout of the room.

Easy All of the target objects and agents are in one single room. In this setting, multiple agents are initialized in random positions.

Hard The target objects and multiple agents are distributed in different rooms, and some agents need to go from one room to another to complete the specified task. This is an extension of the AI2-THOR environment. In this setting, multiple target objects are distributed in different rooms, and multiple agents are initialized in random rooms.

Table 3.1 Dataset splits

	Train	Validation (Unseen)	Test (Unseen)
#Tasks	4025	728	926
#Scenes	80	20	20
#Demonstrations	75260	9465	22310

We train the models only in *Easy* setting and evaluate the performance in both *Easy* and *Hard* settings to assess the generalization ability of models.

3.2.3.3 Evaluation Metrics

We utilize two metrics to compare the performance of all models: success rate (SR), success weight by path length (SPL). If all subtasks are completed successfully under the temporal order constraints, the given task is considered successful:

1. **Success rate (SR)**: SR is the ratio of successful tasks, which is defined as follows:

$$SR = \frac{1}{N_{task}} \sum_{i=1}^{N_{task}} R_i,$$

where $R_i = 1$ if the i-th task is successful, otherwise $R_i = 0$, and N_{task} is the number of tasks.
2. **Success weight by path length (SPL)**: SPL is denoted as follows:

$$SPL = \frac{1}{N_{task}} \sum_{i=1}^{N_{task}} R_i \frac{L_i}{max(D_i, L_i)},$$

where L_i denotes the minimum number of steps for multiple agents to complete the task, and D_i is the actual steps.

3.2.3.4 Comparison of Task Planning

We conduct experiments to evaluate the effectiveness of our task planning module, including task decomposition and task allocation parts.

In the task decomposition part, we are interested in the difference between Fixed Task Decomposition (FTD) and Dynamic Task Decomposition (DTD). The former provides task decomposition results directly from the external knowledge base, while DTD will further dynamically adjust the results using the perception information. In addition, to show the performance gap to the upper bound, we also introduce the Oracle Task Decomposition (Oracle-TD) method, which uses the ground truth of the task decomposition.

In the task allocation part, we first design two baseline methods: For *Random* method, the agents randomly choose the task allocation for execution. For *Init.* method, the agents initially use some simple rules that satisfy the allocation constrictions to form the allocation results and keep them unchanged during the running period. We consider two cases of our proposed task allocation method: *Fixed Task Allocation (FTA)* and *Dynamic Task Allocation (DTA)*: The former calculates the allocation results only once, and the latter dynamically adjusts the allocation using the perception information. Finally, we replace the calculation method in *Fixed Task Allocation (FTA)* and *Dynamic Task Allocation (DTA)* with the ground truth results to form the ceiling methods *Oracle-FTA* and *Oracle-DTA*.

Therefore, there are a total of $3 \times 6 = 18$ combinations considering different task decomposition and allocation ways, and we evaluate them in two types of difficulty setting with three types of tasks, respectively. The results of different types of instructions are shown in Tables 3.2, 3.3, and 3.4, respectively, from which we draw the following conclusions:

1. **Dynamic task decomposition significantly performs better than fixed task decomposition:** The reason is that FTD simply reasons receptacles based on natural language without considering visual perceptions, while DTD dynamically updates the decomposition results according to agents' perception information. An exception appears in Type III, in which the receptacles required for inference in the type *two action and two objects* are relatively simple and can be accurately reasoned directly based on fixed decomposition. Therefore, the success rates of different decomposition models in this type of task are the same. We also observe that the oracle decomposition method Oracle-TD performs best in each type of tasks because it uses the true decomposition results for the subsequent process. This reflects that the task decomposition significantly affects the subsequent task allocation and navigation process, so it has a great influence on the success rate of the whole task. The results also show that there still exists a gap between Oracle-TD and our DTD, which means the embodied task decomposition is challenging and still has large space for further research.
2. **Dynamic task allocation significantly performs better than fixed task allocation:** It is found that the dynamic allocation methods always have advantages

Table 3.2 Comparative results in task planning module in type I

		Task decomposition					
		Oracle-TD		FTD		DTD	
Settings	Task allocation	SR (%)	SPL (%)	SR (%)	SPL (%)	SR (%)	SPL (%)
Easy	Random	0.0	0.0	0.0	0.0	0.0	0.0
	Init	60.3	12.0	14.2	3.1	35.6	7.1
	FTA	77.4	15.9	17.3	4.4	44.9	9.7
	DTA	**78.2**	**16.7**	18.9	**5.1**	**45.1**	**10.3**
	Oracle-FTA	78.2	16.2	17.7	4.4	46.5	9.7
	Oracle-DTA	81.9	18.2	20.0	5.5	49.0	11.3
Hard	Random	0.0	0.0	0.0	0.0	0.0	0.0
	Init	31.6	19.7	11.5	6.5	13.9	7.6
	FTA	46.9	28.2	18.1	11.8	18.6	12.0
	DTA	**50.2**	**28.9**	**19.6**	**12.5**	**22.0**	**13.2**
	Oracle-FTA	55.0	30.7	20.6	13.9	21.5	13.5
	Oracle-DTA	55.5	30.5	21.5	13.4	23.0	13.7

Table 3.3 Comparative results in task planning module in type II

		Task decomposition					
		Oracle-TD		FTD		DTD	
Settings	Task allocation	SR (%)	SPL (%)	SR (%)	SPL (%)	SR (%)	SPL (%)
Easy	Random	0.0	0.0	0.0	0.0	0.0	0.0
	Init	13.7	3.2	2.4	0.8	8.7	2.2
	FTA	26.4	6.0	3.8	0.9	14.7	3.6
	DTA	**29.6**	**7.3**	**4.6**	**1.1**	**17.3**	**4.4**
	Oracle-FTA	31.3	7.5	5.3	1.4	17.3	4.5
	Oracle-DTA	35.6	9.6	5.5	1.4	18.8	5.3
Hard	Random	0.0	0.0	0.0	0.0	0.0	0.0
	Init	6.7	2.3	2.6	0.2	4.0	1.3
	FTA	10.7	3.7	2.7	0.9	8.0	3.2
	DTA	**13.3**	**5.1**	**5.3**	**2.3**	**9.3**	**3.9**
	Oracle-FTA	20.0	6.7	9.3	3.2	14.7	5.3
	Oracle-DTA	21.3	7.0	12.0	4.0	17.3	5.8

Table 3.4 Comparative results in task planning module in type III

		Task decomposition					
		Oracle-TD		FTD		DTD	
Settings	Task allocation	SR(%)	SPL(%)	SR(%)	SPL(%)	SR(%)	SPL(%)
Easy	Random	0.0	0.0	0.0	0.0	0.0	0.0
	Init	22.0	5.7	22.0	5.7	22.0	5.7
	FTA	22.0	6.1	22.0	6.1	22.0	6.1
	DTA	**27.0**	**6.8**	**27.0**	**6.8**	**27.0**	**6.8**
	Oracle-FTA	31.0	7.2	31.0	7.2	31.0	7.2
	Oracle-DTA	35.0	8.9	35.0	8.9	35.0	8.9
Hard	Random	0.0	0.0	0.0	0.0	0.0	0.0
	Init	3.0	0.8	3.0	0.8	3.0	0.8
	FTA	4.5	0.8	4.5	0.8	4.5	0.8
	DTA	**13.4**	**2.9**	**13.4**	**2.9**	**13.4**	**2.9**
	Oracle-FTA	11.9	3.1	11.9	3.1	11.9	3.1
	Oracle-DTA	14.9	3.8	14.9	3.8	14.9	3.8

over the specific fixed ones. Due to the mutual influence of task allocation and navigation, agents perform navigation based on the allocated subtasks, and agents can obtain new distance prediction results with the movement and new visual observations. Dynamic allocation methods can dynamically update the allocation results based on new perceptions at each step, which is more reasonable than the fixed strategy.

3. **The embodied task planning achieves the best results except Oracle:** Agents have less visual perceptions about the scene at the beginning of the task, and then the receptacle reasoning and distance prediction become more accurate with

the continuous interaction and perception with the scene, which can generate more accurate allocation results and improve the accuracy of navigation and task completion. The results demonstrate the importance and effectiveness of the embodied task planning module.

3.2.3.5 Qualitative Results

We show the successful sample of multi-agent task assignment in the easy setting in Fig. 3.4. Agents execute the setting *"Put the cup away"* in the same room. The implicit receptacle is reasoned to be *Cabinet*, and decomposed subtasks are allocated to three agents based on their visual perception at the first step. Then at the 3rd step, Agent 1 and Agent 3 change their subtasks because of their newly obtained visual perceptions that can help them predict the distance to target objects more accurately. At the 5th step, after Agent 2 finishes its subtask, it switches the subtask and completes the whole task with the help of communication messages from Agent 1 and Agent 3.

We also demonstrate the successful sample of multi-agent task assignment in hard settings, which is shown in Fig. 3.5. Agents perform *"Clean the plate and put*

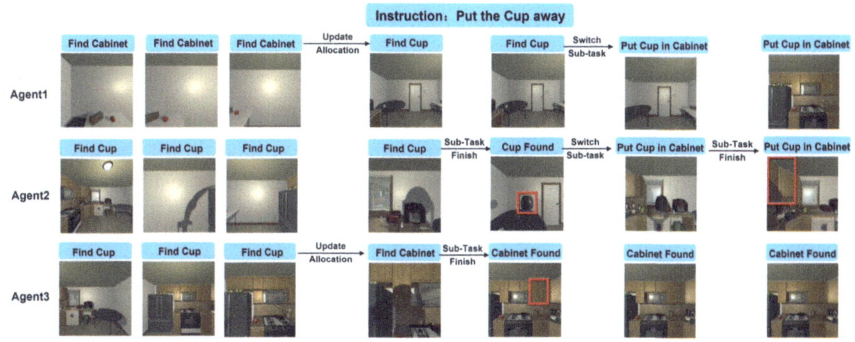

Fig. 3.4 Qualitative results of the *Easy* setting in the simulation environment

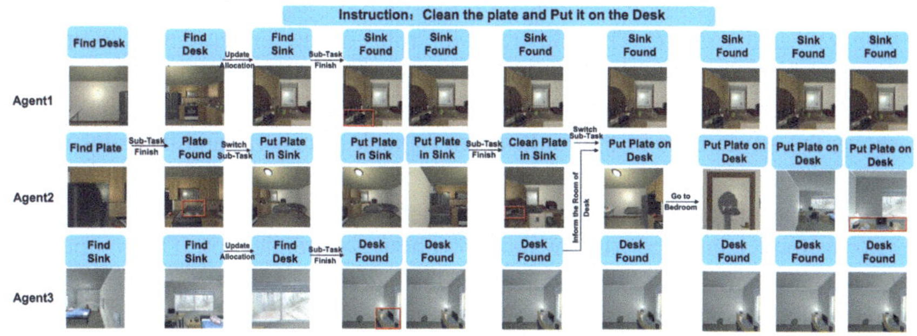

Fig. 3.5 Qualitative results of the *Hard* setting in the simulation environment

it on the desk" in different rooms. At the 3rd step, Agent 1 and Agent 3 dynamically adjust the task allocation results with the updated perceptions of the scene. After Agent 2 finishes its subtasks *Clean the plate in the sink*, it switches to *Put the plate on the desk*, goes into the bedroom with the help of the effective information from Agent 3, and finally completes the task successfully. This successful sample demonstrates that the embodied dynamic task allocation is important and effective in the multi-agent task planning process.

3.2.3.6 Multi-Agent vs. Single-Agent

Before closing this section, we present results to show the advantages of multi-agent collaboration in these scenarios and therefore further verify the roles of the embodied task planning since it is only useful for multi-agent. To this end, we compare the results of our multi-agent task planning framework in three types of tasks with the following methods.

For single-agent cases, we consider three methods. The most typical one is *VSN-SP* in which the scene prior is incorporated into this model except for the RGB observations and the previous action as input [22]. This method reduces to *VSN* if we remove the scene prior part. Further, we set the number of agents to one in our method and form the method *Single-agent*, which does not have the task planning and communication module. The core difference between *Single-agent*, *VSN-SP*, and *VSN* is that *Single-agent* utilizes the subgoal planning method, while *VSN-SP* and *VSN* use the low action generation policy.

For multi-agent cases, we consider two methods. The first one is named *Multi-agent w/o com*, which does not have the communication module for the agents to exchange information during the collaborative navigation, and its rest is the same as our multi-agent framework. The other one *Multi-agent* is our standard and complete version.

To perform a fair comparison between multi-agent and single-agent baselines, we use the oracle task decomposition and allocation results, and the SPL is calculated based on the low-level actions. The comparison results are shown in Table 3.5, from which we make the following observations:

1. Even for the single-agent case, the proposed method *Single-agent* performs better than *VSN* and *VSN-SP*. This illustrates the roles of the semantic map coding for the visual perception and subgoal planning for the action generation.
2. The *Multi-agent* method performs much better than *Single-agent* methods. This shows the advantages of the multi-agent in such a scenario and the roles of the task planning.
3. Type III has higher requirements for collaboration, and our model has significant improvement compared with the model without communication. The comparison results between *Multi-agent* and *Multi-agent w/o Com.* demonstrate the effectiveness of the communication module.

Table 3.5 Comparative results for multi-agent vs. single-agent

		Type I		Type II		Type III	
Settings	Methods	SR (%)	SPL (%)	SR (%)	SPL (%)	SR (%)	SPL (%)
Easy	VSN	13.2	2.9	2.0	0.2	4.0	1.3
	VSN-SP	15.0	3.0	4.0	0.5	9.5	2.5
	Single-agent	46.3	10.4	10.1	2.3	22.0	8.7
	Multi-agent w/o Com.	77.8	16.0	29.8	7.0	26.5	6.0
	Multi-agent	**78.2**	**16.2**	**31.3**	**7.5**	**31.0**	**7.2**
Hard	VSN	8.1	3.8	0.0	0.0	0.0	0.0
	VSN-SP	11.0	5.7	0.0	0.0	1.5	0.2
	Single-agent	23.9	10.6	1.3	0.8	4.5	1.6
	Multi-agent w/o Com.	52.7	26.4	18.7	**8.0**	7.5	1.9
	Multi-agent	**55.0**	**30.7**	**20.0**	6.7	**11.9**	**3.1**

3.3 Decentralized Perception-Action Loop in Homogeneous Embodied Multi-Agent System

In this section, we analyze the decentralized perception-action loop in the homogeneous embodied multi-agent system in the collaborative object search problem. We introduce the embodied collaborative object search problem and discuss the perception-action framework to solve this problem. Then we demonstrate the experiment settings and results in the simulation environment.

3.3.1 *Problem Formulation*

In the multi-agent embodied collaborative object search task, multiple agents aim to find multiple objects collaboratively with egocentric visual observations and the class labels of target objects, which is demonstrated in Fig. 3.6.

In a scene $\mathcal{S}$, there exist multiple objects of different categories which are denoted as $O = \{O^{(1)}, O^{(2)}, \cdots, O^{(K)}\}$, where each $O^{(k)} (k = 1, 2, \cdots, K)$ represents a certain type of the objects, and K is the number of existing object types in $\mathcal{S}$. There are N collaborative homogeneous agents $A^{(1)}, A^{(2)}, \cdots, A^{(N)}$. The target objects in the task are denoted as the set $\mathcal{T} = \{T^{(1)}, T^{(2)}, \cdots, T^{(M)}\}$, in which M represents the number of target object categories, and $\mathcal{T} \subseteq O$ is satisfied. The goal of these collaborative agents is to find objects from all the categories in $\mathcal{T}$, and if there are multiple instances of a certain object type in the scene, agents only need to find one of them. In particular, multiple target objects are not assigned to specific agents, and agents need to decide which object to navigate based on their observations.

At every timestep t, the i-th agent $A^{(i)}$ can obtain the egocentric visual observation $v_t^{(i)}$, the communication information from other agents $c_t^{(i)}$, the history state information $h_{t-1}^{(i)}$, as well as the class labels of targets $\mathcal{T}$. Every agent $A^{(i)}$

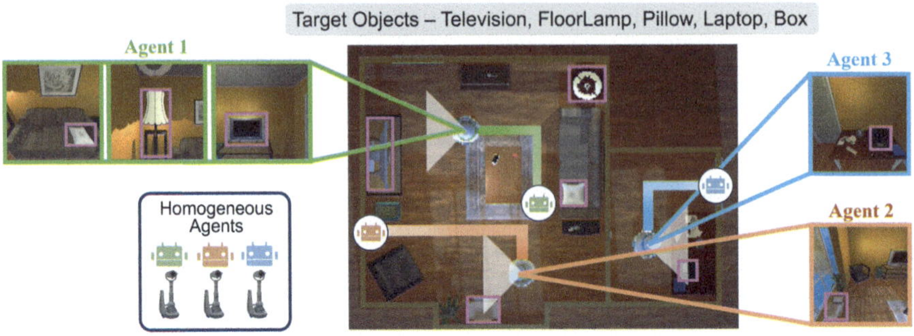

Fig. 3.6 An overview of the decentralized homogeneous multi-agent collaborative object search task

learns a proper action strategy $\pi^{(i)}$ to generate action $a_t^{(i)}$ at timestep t, where $a_t^{(i)} = \pi^{(i)}(v_t^{(i)}, c_t^{(i)}, h_{t-1}^{(i)}, \mathcal{T})$ and the action $a_t^{(i)}$ is chosen from the action set $\mathcal{A}$. Specifically, the action space is defined as $\mathcal{A} = \{MoveAhead,\ RotateRight,\ RotateLeft,\ LookUp,\ LookDown,\ Found,\ Done\}$. $MoveAhead$ means moving straight along the orientation for 0.25 m. $RotateRight$ and $RotateLeft$ indicate turning right or left for 90°, respectively. $LookUp$ and $LookDown$ mean turning the view up or down for 30°, respectively. $Found$ means that the agent has found a target object that is within the view of the agent, and the distance between the agent and the target is less than 1.0 m. $Done$ means that the agent has finished all its actions and stops. After each agent $A^{(i)}$ executes their action $a_t^{(i)}$, they obtain the new state including visual observation $v_{t+1}^{(i)}$, communication information $c_{t+1}^{(i)}$, and history information $h_t^{(i)}$. Agents then continue to generate actions with $\pi^{(i)}$ in the above manner until all agents perform $Done$ action or the maximum number of steps is reached. The objective of this task is to learn the optimal decision policy $\pi^{(i)}$ for every agent $A^{(i)}$ to collaboratively find all target objects with as few steps as possible.

3.3.2 Perception-Action Framework

The proposed decentralized perception-action framework for this collaborative object search problem is composed of five main modules: the semantic mapping module, the key objects mapping module, the communication module, the feature fusion module, and the action decision module. The overview of the proposed model structure is demonstrated in Fig. 3.7. The semantic mapping module generates a top-down semantic map with RGB and depth images of egocentric visual observations. The key objects mapping module marks the location of key objects with the scene prior knowledge. The communication module is responsible for transmitting and receiving information from other agents. The feature fusion module extracts the

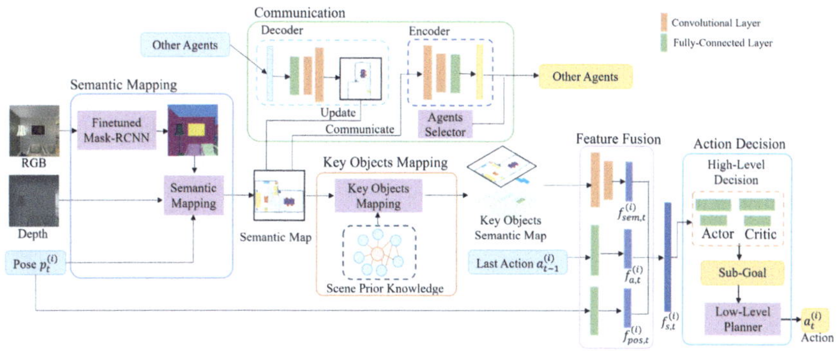

Fig. 3.7 An overview of the proposed model structure

features of target objects and history states. Then the action decision module decides the next action the agent should perform.

3.3.2.1 Semantic Mapping

The semantic mapping module utilizes a pretrained Mask-RCNN [5] to obtain the semantic segmentation results. Then, the semantic segmentation results, the depth image, and the pose of the agent are used to project each pixel with a specific object category into the 3D world coordinates to obtain the 3D voxel semantic map for each object category. We sum along the height dimension of the voxel map of each category and concatenate these maps of each object category to get a top-down semantic map of size $L \times W \times (K_{total} + 2)$, where $L \times W$ represents the size of the map, and K_{total} is the number of object categories. The first K_{total} channels denote the area of each category, and the last two channels represent the occupied and explored area, respectively. Each element in the map denotes a grid of size $5\,\text{cm} \times 5\,\text{cm}$ in simulation scenes.

To generate the semantic map more accurately, we fine-tune the pretrained Mask-RCNN in our training scenes. Specifically, we use the obtained egocentric RGB images and ground truth segmentation results from AI2-THOR and RoboTHOR platforms to fine-tune the Mask-RCNN model for object categories in our experimental scenes. The fine-tuned model is then utilized in the semantic mapping process.

3.3.2.2 Key Objects Mapping

The key objects mapping module utilizes the scene prior knowledge to find the key objects with the current visual observation, and the key objects map is generated.

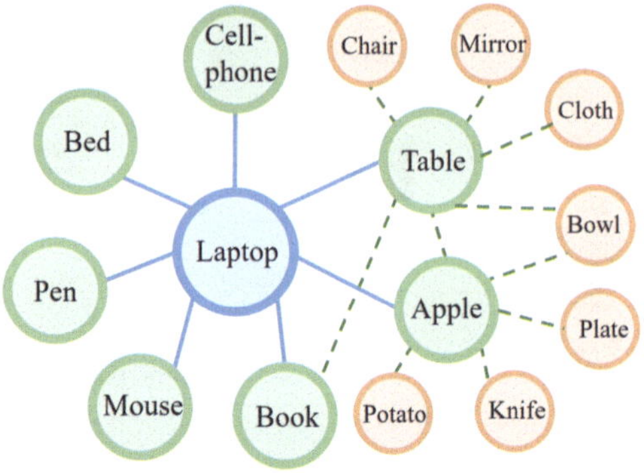

Fig. 3.8 A part of the scene prior knowledge. The solid lines represent the relationships between Laptop and other objects. The dotted lines denote several relationships between Table, Apple, and other objects

Scene Prior Knowledge We build the scene prior knowledge graph using the method similar to Ref. [22]. The scene prior knowledge graph is represented as $\mathcal{G} = (Q, E)$, in which each node $q \in Q$ denotes a class of object, and each edge $e \in E$ represents a spatial relationship between two specific object categories. In this task, we extract the spatial relationships for all object categories in AI2-THOR and RoboTHOR from the Visual Genome [11] to form the scene prior knowledge graph. For example, the laptop is always next to the cell phone, and then there exists an edge between the node Laptop and the node Cell phone. A part of the graph is shown in Fig. 3.8.

Key Objects Extraction The key objects include the target objects and the objects that have common spatial relationships with target objects. The key objects map is constructed by marking key objects based on the semantic map, and it is of size $L \times W \times 2$, in which the first layer marks the target objects and the second layer marks the other key objects having relationships with target objects. This module judges whether there exist key objects using the scene prior knowledge graph in the current visual observation. If no key object exists, the map of the key objects would be denoted as a matrix with all zero elements. If there exist key objects within the view, the module would convert the location of the key objects in visual images into the location in a top-down map with a method similar to the semantic mapping process and then mark the corresponding pixels in the map. Meanwhile, the model would mark all key object instances of the same categories based on the built semantic map. These marked key objects can help agents move to the position as close to target objects as possible.

3.3.2.3 Communication

The communication module includes two parts: the information processing part and the agent selection part. The information processing part is utilized to encode and decode the relevant information. The agent selection part chooses which agent the encoded information is transmitted to according to the specific regulation.

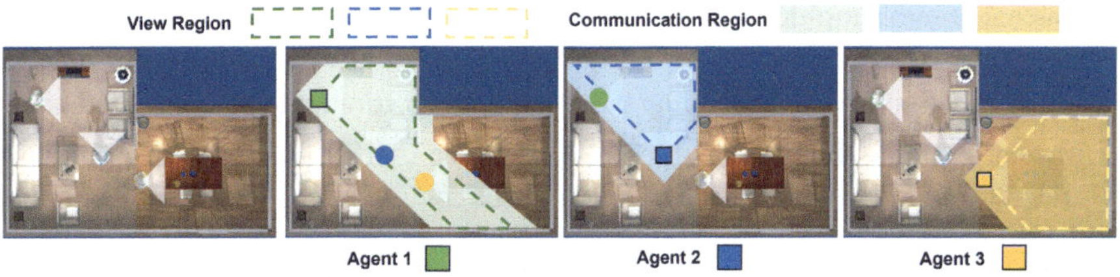

Fig. 3.9 The communication area of three agents. The painted area denotes the visible area, and the dotted area denotes the whole communication area. The colored circles indicate agents in the communication area. At this timestep, Agent 1 sends the message to Agent 2 and Agent 3. Agent 2 sends the message to Agent 1. Agent 3 has no communication

Information Processing In order to decrease the amount of data transmission during communication and reduce the required communication bandwidth, the information processing part, composed of an encoder and a decoder, is utilized to encode the semantic information into a vector to reduce the amount of communication information. The encoder consists of two convolutional layers and a fully connected layer. It can encode the constructed semantic map into a one-dimensional vector. The decoder is composed of a fully connected layer and two upsampling convolutional layers. It is utilized to decode the received communication vector into a semantic matrix. We train the encoder and decoder in advance. The original semantic maps are fed into the encoder and the decoder successively to generate recovered semantic maps, and we use the original semantic maps as supervision to train the encoder and decoder. The difference between the original and the recovered semantic maps should be as small as possible.

Agent Selection The agent selection part helps agents choose the specific agents to transmit the communication information before each round of communication. Every agent selects the agents who are more likely to effectively utilize the communication information to communicate, which saves communication bandwidth compared with directly broadcasting to all agents. In this task, the communication area of each agent in every timestep is defined as the area currently visible to the agent and the area within 1.0 meter from the boundary of the visible area, as is shown in Fig. 3.9. In each round of communication, agents only send the encoded information to other agents in their communication area, as agents in the communication area are much more likely to use the semantic map information in the next few steps.

3.3.2.4 Feature Fusion

The feature fusion module takes the built semantic map, the key objects map, the current pose, and the last performed action as input to extract the feature of the current state of the agent. At timestep t, the semantic map and key objects map are concatenated into a top-down key objects semantic map of size $L \times W \times (K_{total}+4)$. We use the CNN network with two convolutional layers to generate the semantic feature $f_{sem,t}^{(i)}$ of the map, preserving the spatial information. A fully connected layer is utilized to obtain the pose feature $f_{pos,t}^{(i)}$ from the current pose $p_t^{(i)}$, mapping the pose feature to the same feature space of the semantic feature. Meanwhile, another fully connected layer is used to project the last action vector $a_{t-1}^{(i)}$ to the same feature space, generating the action feature $f_{a,t}^{(i)}$. Then, we concatenate feature $f_{sem,t}^{(i)}$, $f_{pos,t}^{(i)}$, and $f_{a,t}^{(i)}$ to get the integral state feature $f_{s,t}^{(i)}$.

3.3.2.5 Action Decision

The action decision module takes the state feature $f_{s,t}^{(i)}$ as input and decides the next action $a_t^{(i)}$ to perform for the agent $A^{(i)}$ at timestep t. The action decision module is a hierarchical decision module, which is composed of a high-level decision part and a low-level decision part.

High-Level Decision The high-level decision part decides a subgoal $sg_t^{(i)} = (sgx_t^{(i)}, sgy_t^{(i)})$ on the top-down map for agent $A^{(i)}$ based on the state feature $f_{s,t}^{(i)}$. The decision part is trained with PPO [16] reinforcement learning algorithm, which includes an actor and a critic composed of fully connected layers. It generates the subgoal every d steps (in our experiments $d = 10$). If the number of interval steps d is reached or the agent reaches the subgoal with the low-level decision part within d steps, the high-level decision part would replan the subgoal. The reward of this strategy is

$$\begin{aligned}
R_u &= \alpha \Delta dis_{to}(sg_u, pos_{u-1}) + \beta \Delta dis_{ko}(sg_u, pos_{u-1}) \\
\Delta dis_{to}(sg_u, pos_{u-1}) &= dis_{to}(pos_{u-1}) - dis_{to}(sg_u) \\
\Delta dis_{ko}(sg_u, pos_{u-1}) &= dis_{ko}(pos_{u-1}) - dis_{ko}(sg_u),
\end{aligned} \tag{3.5}$$

where $dis_{to}(sg_u)$, $dis_{to}(pos_{u-1})$ represent the Euclidean distance from the subgoal sg_u, the current position pos_{u-1} to one nearest target object, respectively, and $dis_{ko}(sg_u)$, $dis_{ko}(pos_{u-1})$ denote the Euclidean distance from sg_u, pos_{u-1} to one nearest key object except for target objects, respectively. $\Delta dis_{to}(sg_u, pos_{u-1})$ is the decrease in distance from the next subgoal to the nearest target object, and if the value is greater than 0, the next subgoal is closer to the target object compared with the current position. The same is true for Δdis_{ko}. α and β are hyper-parameters to control the calculation of the reward function, and we set $\alpha = 0.7, \beta = 0.3$ in this task. This reward function denotes that both the distances from the subgoal to the nearest target object and the nearest key object are taken into consideration, incorporating the effect of the scene prior knowledge.

Low-Level Decision The low-level decision part decides the specific actions for the agent using the modified breadth-first search [2] shortest path algorithm to reach the subgoal generated by the high-level decision part. The low-level decision module generates the action based on the updated semantic map at every step, in which the occupied channel denotes the obstacle position and the explored channel represents the region that has been walked through.

3.3.3 Simulation Experiments

3.3.3.1 Dataset

We conduct experiments on two platforms: AI2-THOR [10] and RoboTHOR [3]. AI2-THOR includes 120 indoor scenes from four room types: *Kitchens*, *Living rooms*, *Bedrooms*, and *Bathrooms*. There are 80 training scenes and 40 testing scenes totally in AI2-THOR. RoboTHOR contains a more complicated room structure with more wall blocks, and there are a total of 60 training scenes and 15 testing scenes. Training scenes are different from testing scenes, and we evaluate our model in unseen testing scenes. In every scene, we select M ($M = 1, 2, 3, 4, 5$, respectively) target objects to form the navigation task and ensure that every target has at least one instance. In AI2-THOR, we generate 7900 training tasks and 4000 testing tasks. In RoboTHOR, we generate 5940 training tasks and 1485 testing tasks.

3.3.3.2 Implementation Details

In our experiments, the number of target objects is set to be $M = 1, 2, 3, 4, 5$, and the number of collaborative agents is set to be $N = 1, 2, 3, 4, 5$, respectively. In every task, N agents are initialized at random reachable positions. We set $N = 1, 2, 3, 4, 5$ respectively to perform multi-agent navigation in all tasks.

We train the encoder and decoder in several training scenes in advance. Twenty training scenes are chosen from the whole training set. We put an agent in every reachable pose in each scene to get the RGB and depth images. Then 2D semantic maps $\{Sem_1, Sem_2, \cdots, Sem_{cn}\}$ are generated with the method mentioned in Section IV.A, where cn is the number of the generated semantic maps. Then we feed the semantic maps into the encoder $\boldsymbol{En}$ and the decoder $\boldsymbol{De}$ successively to get the recovered maps results $Sre_i = \boldsymbol{De}(\boldsymbol{En}(Sem_i))$, where $i = 1, 2, \cdots, cn$ and Sre_i represents the recovered semantic maps. The loss function of this training process is defined as the pixel-wise distance between original semantic maps and recovered semantic maps,

$$Loss_{com} = \frac{1}{nc} \sum_{i=1}^{nc} Dis(Sre_i, Sem_i), \tag{3.6}$$

where Dis denotes the pixel-wise distance between Sre_i and Sem_i, and the distance should be as small as possible.

We train the high-level decision module with the PPO algorithm. We use Adam optimizer during training with a learning rate of 0.0001. We set the discount factor $\gamma = 0.99$.

3.3.3.3 Baselines

We compare our models with the following baselines to evaluate the effectiveness of the proposed model:

1. **Random**: N agents randomly perform actions from the action space $\mathcal{A}$ at each timestep.
2. **MapFBE**: N agents use classical mapping and classical frontier-based exploration (FBE) algorithm [21]. If the agent detects that target objects exist in the visual observation using Mask-RCNN, the agent would navigate toward the detected target objects based on a path planner.
3. **RGB-D**: This model extracts the visual feature of the RGB-D input using ResNet18 [6] and decides the action based on A3C [14] without communication. It directly extends the single-agent model to the multi-agent task.
4. **RGB-D+SP**: This model uses GCN to introduce the scene prior knowledge to *RGB-D+RL*, and it extends the single-agent navigation model with scene priors [22] to multiple agents without communication.
5. **RGB-D+SS**: This model also extracts the semantic segmentation (SS) feature of the RGB-D input using Mask-RCNN. The rest is the same as *RGB-D+RL*.
6. **Central**: In this model, there is a central agent that has access to observations of all agents and makes decisions for agents based on the whole information. The other parts of this model are the same as our model.
7. **Cen w/o SP**: This model does not utilize the scene prior knowledge, and the rest is the same as *Central*.
8. **CordSyn [8]**: We utilize the structure of the decision policy proposed in [8], which is a decentralized model without scene prior knowledge, and modify it to apply to multi-agent navigation.
9. **ColNav [19]**: We utilize the structure of the decision module proposed in [19] and modify it to adapt to our experimental settings, which is a decentralized multi-agent learning model without scene prior knowledge.

3.3.3.4 Evaluation Metrics

We use three main metrics to evaluate the performance of our model: success rate (SR), success weighted by path length (SPL) proposed in [1], and efficiency improvement (EI). A task is considered successful if all targets have been found successfully, otherwise the task fails:

1. **Success rate (SR)**: SR is defined as follows:

$$SR = \frac{1}{N_{task}} \sum_{i=1}^{N_{task}} R_i,$$

where $R_i = 1$ if the i-th navigation task is successful, otherwise $R_i = 0$, and N_{task} is the number of tasks.

2. **Success weighted by path length (SPL)**: SPL is defined as follows:

$$SPL = \frac{1}{N_{task}} \sum_{i=1}^{N_{task}} R_i \frac{L_i}{\max(D_i, L_i)},$$

where L_i denotes the minimum number of steps to find all targets for N agents in the i-th task, and D_i is the actual steps for N agents to complete the task. Since N agents navigate to targets simultaneously, D_i is the number of steps of the last agent performing *Done*.

3. **Efficiency Improvement (EI)**: EI is calculated as follows:

$$EI = \frac{1}{N_{suc}} \sum_{i=1}^{N_{task}} B_i \frac{E_i - D_i}{E_i},$$

where $B_i = 1$ if the i-th task is successful with both N agents and a single agent, otherwise $B_i = 0$. N_{suc} is the number of tasks that are successful with both the current N agents and a single-agent. E_i denotes the number of steps a single agent takes to complete the i-th task. The value of EI denotes the improved percentage of efficiency of multiple agents compared with an individual agent in the same task.

3.3.3.5 Quantitative Results

We compare the results of the proposed model with other baselines in unseen testing scenes with both seen objects and unseen objects. The *Seen Objects* setting means that target objects in testing scenes have appeared as targets in training tasks, while the *Unseen Objects* setting means target objects in testing scenes have not appeared as targets during training. The quantitative performances in AI2-THOR and RoboTHOR are illustrated in Tables 3.6 and 3.7, respectively. *Central** denotes that *Central* performs the best with different values of N, since it can obtain all observations of all agents when making action decisions, which is only regarded as the upper bound for reference. The results of *Central* are marked in italic type, and the best results except for *Central* in each column are marked in bold. We can obtain the following results:

Table 3.6 Quantitative results in AI2-THOR

Setting	Methods	$N = 1$			$N = 2$			$N = 3$			$N = 4$			$N = 5$		
		SR	SPL	EI	SR	SPL	EI	SR	SPL	EI	SR	SPL	EI	SR	SPL	EI
Seen objects	Random	8.02	3.12	–	8.31	3.22	10.01	8.33	3.35	12.10	8.51	3.76	12.25	8.55	3.79	12.27
	MapFBE	28.23	14.71	–	28.39	14.92	28.37	28.90	15.23	37.35	29.33	16.66	37.62	29.89	16.53	37.98
	RGB-D	23.52	13.76	–	23.75	13.98	25.72	24.33	14.05	35.42	24.60	14.32	35.88	24.75	14.55	37.36
	RGB-D+SP	25.80	14.72	–	26.29	14.90	28.35	27.55	15.19	37.30	28.40	16.35	37.55	28.62	16.50	37.90
	RGB-D+SS	23.60	13.81	–	24.03	14.23	26.05	24.52	14.65	35.72	24.88	14.69	36.05	25.12	15.02	37.50
	CordSyn	35.73	14.88	–	36.25	15.30	28.56	36.80	16.61	38.81	37.02	17.03	39.23	37.15	17.82	39.35
	ColNav	33.23	14.77	–	34.99	15.01	28.40	35.31	15.36	38.05	36.25	16.88	38.01	36.72	17.23	38.33
	Cen w/o SP	41.00	15.10	–	43.17	15.56	29.07	44.03	18.01	40.03	44.28	18.32	40.31	44.50	18.39	40.36
	Ours	**45.19**	**18.25**	–	**47.02**	**19.13**	**31.01**	**47.60**	**21.99**	**41.33**	**47.77**	**23.25**	**43.55**	**48.02**	**23.30**	**43.60**
	*Central**	*45.19*	*18.25*	–	*47.93*	*19.66*	*32.10*	*48.16*	*22.33*	*42.00*	*48.71*	*23.56*	*43.59*	*48.90*	*23.69*	*43.75*
Unseen objects	Random	8.00	3.13	–	8.28	3.20	9.96	8.31	3.32	12.05	8.45	3.76	12.09	8.53	3.70	12.12
	MapFBE	26.02	11.72	–	26.23	12.55	26.21	26.85	12.79	34.98	27.63	13.01	35.23	27.60	13.75	35.60
	RGB-D	21.03	9.89	–	21.25	10.23	23.15	21.90	11.01	33.21	22.01	11.95	33.65	22.43	12.15	35.21
	RGB-D+SP	23.65	11.53	–	23.95	11.79	26.03	24.05	12.06	34.77	24.65	12.55	35.02	24.89	13.01	35.35
	RGB-D+SS	22.05	10.15	–	22.66	10.76	24.15	23.15	11.37	33.66	23.93	12.15	34.01	24.22	12.37	35.28
	CordSyn	32.55	12.03	–	33.19	12.88	27.35	34.12	13.27	35.20	35.33	14.85	36.55	35.63	15.05	36.61
	ColNav	30.69	11.85	–	31.26	12.63	26.95	32.81	13.03	35.02	33.32	13.95	35.98	33.50	14.10	36.07
	Cen w/o SP	39.73	13.06	–	40.20	13.75	28.35	41.88	15.12	39.12	43.01	17.00	40.17	43.29	17.22	40.26
	Ours	**43.81**	**16.35**	–	**43.95**	**16.55**	**30.00**	**44.56**	**17.25**	**39.05**	**45.25**	**17.93**	**39.98**	**46.17**	**18.07**	**40.02**
	*Central**	*43.81*	*16.35*	–	*43.98*	*17.01*	*30.75*	*45.15*	*18.85*	*40.33*	*46.33*	*19.60*	*41.35*	*46.72*	*19.77*	*41.66*

[1] All values of SR, SPL, and EI are percentages in this table.

Table 3.7 Quantitative results in RoboTHOR

		$N=1$			$N=2$			$N=3$			$N=4$			$N=5$		
Setting	Methods	SR	SPL	EI	SR	SPL	EI	SR	SPL	EI	SR	SPL	EI	SR	SPL	EI
Seen objects	Random	6.75	2.83	–	6.83	2.91	9.35	6.88	3.01	11.35	7.12	3.09	11.39	7.30	3.55	11.88
	MappFBE	25.12	11.53	–	25.33	11.80	25.82	25.60	12.89	36.56	26.05	13.03	36.90	26.59	13.22	37.55
	RGB-D	21.23	10.40	–	21.98	10.65	23.33	22.12	11.20	34.10	22.30	12.10	34.35	22.55	12.39	37.02
	RGB-D+SP	22.75	11.09	–	23.01	11.75	25.80	23.45	12.80	35.30	24.36	12.96	35.98	25.05	13.21	37.20
	RGB-D+SS	21.56	10.60	–	22.12	10.96	23.60	22.71	11.59	34.39	22.90	12.56	34.78	23.02	12.65	37.16
	CordSyn	31.87	11.77	–	32.22	12.30	26.65	34.05	13.65	38.02	35.11	14.63	39.01	35.65	14.88	39.20
	ColNav	30.56	11.66	–	31.12	11.98	26.05	32.75	13.06	36.70	33.69	13.56	37.25	34.33	13.85	37.77
	Cen w/o SP	36.15	12.60	–	37.76	12.99	28.88	38.95	15.03	39.95	39.16	15.25	40.35	39.17	15.30	40.39
	Ours	**40.85**	**13.95**	–	**42.20**	**14.17**	**30.55**	**42.33**	**17.75**	**41.55**	**42.52**	**19.06**	**43.62**	**43.01**	**20.23**	**43.65**
	*Central**	*40.85*	*13.95*	–	*42.76*	*15.08*	*31.63*	*42.85*	*18.05*	*42.05*	*42.91*	*19.88*	*43.88*	*43.15*	*20.33*	*43.93*
Unseen objects	Random	6.57	2.80	–	6.83	2.88	9.25	6.85	2.97	11.30	7.10	3.00	11.36	7.22	3.32	11.55
	MappFBE	23.03	9.23	–	23.56	10.06	23.35	24.06	10.32	34.20	24.36	10.83	34.72	24.51	11.65	35.00
	RGB-D	17.98	7.23	–	18.05	7.53	20.01	18.67	8.23	31.02	19.03	8.89	31.60	19.95	9.76	33.23
	RGB-D+SP	19.56	8.35	–	20.05	9.35	21.95	20.56	10.21	32.05	21.01	10.76	32.39	21.65	11.40	34.05
	RGB-D+SS	18.10	7.35	–	18.95	8.06	20.75	19.23	8.90	31.76	20.03	9.35	31.96	20.25	10.33	33.91
	CordSyn	28.55	10.03	–	29.12	11.31	24.85	31.56	12.05	36.77	33.35	12.90	37.50	33.62	13.05	37.89
	ColNav	26.95	9.50	–	27.33	10.65	23.88	28.68	11.39	35.85	30.05	12.05	36.69	30.96	12.33	37.15
	Cen w/o SP	33.49	11.88	–	34.88	12.05	26.10	35.23	13.15	37.53	36.00	14.07	39.42	36.23	14.57	39.55
	Ours	**38.66**	**12.12**	–	**39.89**	**12.65**	**29.36**	**40.88**	**14.96**	**39.33**	**41.15**	**15.50**	**40.85**	**41.33**	**15.72**	**40.99**
	*Central**	*38.66*	*12.12*	–	*40.19*	*13.30*	*30.35*	*41.33*	*15.21*	*40.25*	*41.89*	*16.33*	*41.25*	*42.07*	*16.93*	*41.60*

[1] All values of SR, SPL, and EI are percentages in this table.

1. Our model performs better than other baselines except for *Central*. The poor performances of *RGB-D+RL*, *RGB-D+SP.+RL*, and *RGB-D+SS+RL*, which are directly extended from single-agent models, indicate that the multi-agent navigation task is challenging. The differences between our model and *Central* in SR, SPL, and EI are the smallest, denoting that the proposed multi-agent navigation model is reasonable.
2. Our model improves SR, SPL, and EI with the increasing number of agents, indicating that the encoded communication messages and the agent selection part of the communication module are effective. Meanwhile, the improvement of SR is not as obvious as SPL and EI. As fewer agents take more action steps to make up for their unreasonable decisions, the SR of navigation will not decrease much, while SPL and EI will decrease significantly.
3. Models with scene prior knowledge perform better than models of the same structure without scene prior knowledge, which demonstrates that the scene priors can provide effective navigation information to help agents find a shorter path to target objects.
4. Comparing results on seen and unseen objects, we can find that our model has better generalization ability compared with other baselines. The scene prior knowledge and the semantic map enable agents to infer the locations of target objects where they are more likely to appear, which are beneficial for agents to navigate to targets that have not been seen during training.

3.3.3.6 Qualitative Results

We show some qualitative results in AI2-THOR and RoboTHOR, respectively, in Fig. 3.10. In these samples, the number of agents N is less than the number of target objects M, and agents collaborate with others to successfully find all target objects. Agents do not perform *Done* immediately when finding a certain target object, and they would judge whether they need to continue to navigate to other targets until all targets are found. Objects that agents see at first may not be close to them, and the object closer to the agent may not be in the current field of view. Agents can obtain information from others through communication and navigate to the closer object with as few steps as possible. Meanwhile, multiple agents can significantly improve the navigation efficiency compared with the single agent.

3.3.3.7 Ablation Experiment

We conduct the ablation experiment to evaluate the effect of different modules of our model in AI2-THOR. We remove the semantic mapping module to form the model *w/o SeMap*, which takes RGB-D images as input and utilizes Mask-RCNN to extract semantic segmentation features without building the semantic map. The model *w/o SP* does not use the scene prior knowledge to build the key objects map. We remove the hierarchical decision module to form the model *w/o HieDec*, in

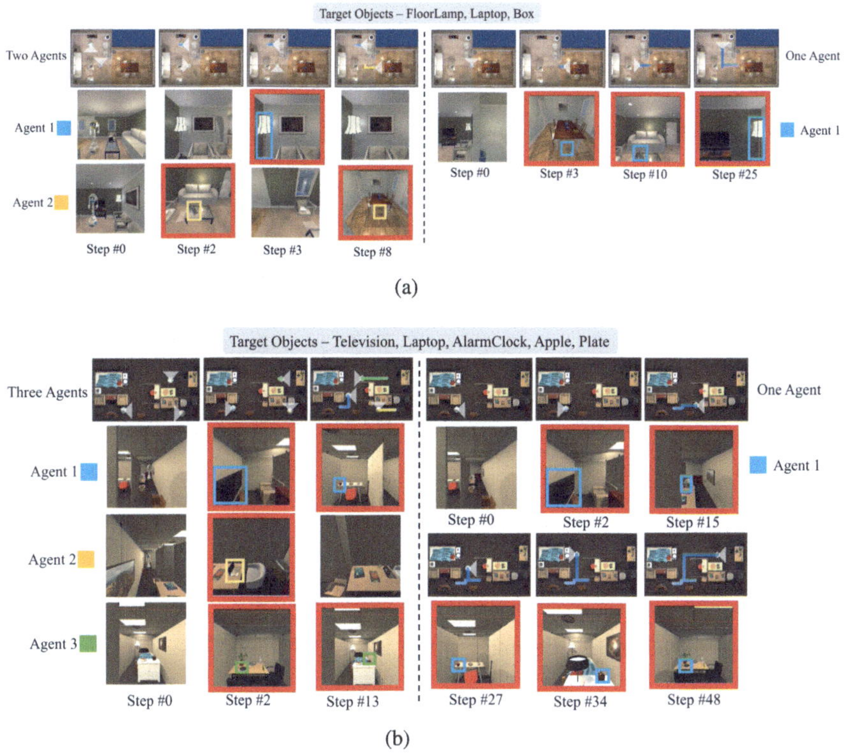

Fig. 3.10 Qualitative results of multi-agent visual semantic navigation. (**a**) shows the navigation process of two agents compared with one agent in *FloorPlan223* in AI2-THOR. (**b**) shows the navigation process of three agents compared with one agent in *FloorPlan_Val1_4* in RoboTHOR. The navigation results of multiple agents are displayed on the left, and the navigation results of a single agent are displayed on the right. Navigation with multiple agents significantly improves the efficiency compared with the single agent, which demonstrates the effectiveness of the proposed model. (**a**) Navigation process of two agents compared with one agent in AI2-THOR. (**b**) Navigation process of three agents compared with one agent in RoboTHOR

which the PPO algorithm is trained to directly generate specific actions from $\mathcal{A}$ instead of generating subgoals in the high-level decision. The model *w/o Comm* removes the communication module.

The results of the ablation experiment are shown in Table 3.8, which indicate that the semantic mapping, scene prior knowledge, hierarchical decision, and communication modules are beneficial for multi-agent navigation. The performance of *w/o SeMap* decreases the most, indicating that the semantic map plays an important role in exploring and understanding scenes. The results of *w/o SP* illustrate that scene prior knowledge is beneficial for multiple agents to find shorter paths to navigate to targets. The performance of *w/o HieDec* demonstrates the effectiveness of the hierarchical decision strategy, and directly predicting specific actions may cause

Table 3.8 Results of the ablation experiment in AI2-THOR

		$N = 1$			$N = 2$			$N = 3$			$N = 4$			$N = 5$		
Setting	Methods	SR	SPL	EI	SR	SPL	EI	SR	SPL	EI	SR	SPL	EI	SR	SPL	EI
Seen objects	w/o SeMap	30.33	12.50	–	31.85	13.39	23.33	33.65	15.06	25.68	34.09	15.25	27.30	34.21	15.30	27.52
	w/o SP	41.00	15.10	–	42.85	15.21	28.55	43.25	17.15	39.02	43.31	18.01	39.95	43.50	18.23	40.01
	w/o HieDec	38.88	14.76	–	40.33	15.09	28.08	42.87	17.03	39.01	43.22	17.75	39.66	43.39	18.00	39.63
	w/o Comm	45.19	18.25	–	42.93	15.61	28.60	43.30	17.31	39.20	44.09	18.35	40.06	43.57	18.45	40.23
	Ours	**45.19**	**18.25**	–	**47.02**	**19.13**	**31.01**	**47.60**	**21.99**	**41.33**	**47.77**	**23.25**	**43.55**	**48.02**	**23.30**	**43.60**
Unseen objects	w/o SeMap	23.50	7.21	–	24.39	8.01	20.12	26.02	10.23	22.10	27.25	11.60	25.11	27.40	11.88	25.60
	w/o SP	39.73	13.06	–	39.33	13.25	27.10	41.16	14.88	38.60	42.59	15.98	38.89	42.70	16.10	39.05
	w/o HieDec	36.50	12.12	–	38.00	13.10	26.56	40.01	14.66	38.55	42.39	15.66	38.80	42.60	15.73	38.91
	w/o Comm	43.81	16.35	–	39.45	13.52	27.35	41.29	14.93	38.73	42.70	16.13	38.93	42.78	16.27	39.21
	Ours	**43.81**	**16.35**	–	**43.95**	**16.55**	**30.00**	**44.56**	**17.25**	**39.05**	**45.25**	**17.93**	**39.98**	**46.17**	**18.07**	**40.02**

[1] All values of SR, SPL, and EI are percentages in this table.

more mistakes in the multi-agent setting. The results of *w/o Comm* denote that communication can help multiple agents navigate to targets in fewer steps.

3.4 Perception-Action Loop in Heterogeneous Embodied Multi-Agent System

In this section, we analyze the perception-action loop in heterogeneous embodied multi-agent system in the embodied multi-agent room tidying-up task. We introduce the multi-agent room tidying-up task and discuss the perception-action framework to solve this task. Then we demonstrate the experiment settings and results in the simulation environment.

3.4.1 Problem Formulation

For the heterogeneous multi-agent collaboration tasks, we consider N heterogeneous agents $A^{(1)}, A^{(2)}, \cdots, A^{(N)}$ in the scene. Each agent $A^{(i)}$ has a capability representation $B^{(i)} = (P^{(i)}, E^{(i)}, H^{(i)})$, which consists of the perception ability $P^{(i)}$, the action ability $E^{(i)}$, and the agent morphology $H^{(i)}$. Specifically, in the heterogeneous multi-agent room tidying-up task, each agent has the same visual perception ability. The action ability $E^{(i)}$ can be indicated as $E^{(i)} = (Nav^{(i)}, Mani^{(i)})$, where $Nav^{(i)}, Mani^{(i)}$ denote whether the agent has the ability of navigation and manipulation, respectively. If the agent has the corresponding ability, its value is 1. Otherwise, the value is 0. The agent morphology $H^{(i)}$ considers different heights of agents and is represented as $H^{(i)} = (Hei^{(i)})$, where $Hei^{(i)} = 1$ denotes the height of the agent is high, roughly 0.9 meters, and $Hei^{(i)} = 0$ denotes the height is low, roughly 0.22 meters. N heterogeneous agents have different $(E^{(i)}, H^{(i)})$, and different action spaces $I^{(i)}$. If $Nav^{(i)} = 1$ and $Mani^{(i)} = 0$, $I^{(i)} = I_{nav}$. If $Nav^{(i)} = 1$ and $Mani^{(i)} = 1$, $I^{(i)} = I_{nav} \cup I_{mani}$, where $I_{nav} = \{MoveAhead, MoveRight, MoveLeft, RotateRight, RotateLeft, LookUp, LookDown, Stop\}$ denotes the navigation action set and $I_{mani} = \{PickUp, PutDown, Drop\}$ represents the manipulation actions.

There exist M objects in the house which are denoted as $O = \{o_1, o_2, \cdots, o_M\}$. The subset $O_{pick} \subseteq O$ denotes the set of objects which can be picked up by agents. The subset $O_{recep} \subseteq O$ represents the set of objects which can be used as a receptacle, and other objects can be placed on their surface or inside. The location of the j-th object is represented as (o_j, p_j, r_j), where $p_j \in O_{recep}$, and o_j is placed in p_j in the room r_j. The environment provides a discriminator D to judge whether the object is in a reasonable location. If $D(o_j, p_j, r_j) = True$, o_j is in a reasonable location, otherwise, o_j is in an unreasonable location. There exist k objects in unreasonable locations in the house, namely misplaced objects, and agents do not know the value of k. At the t-th step, the agent $A^{(i)}$ obtains the

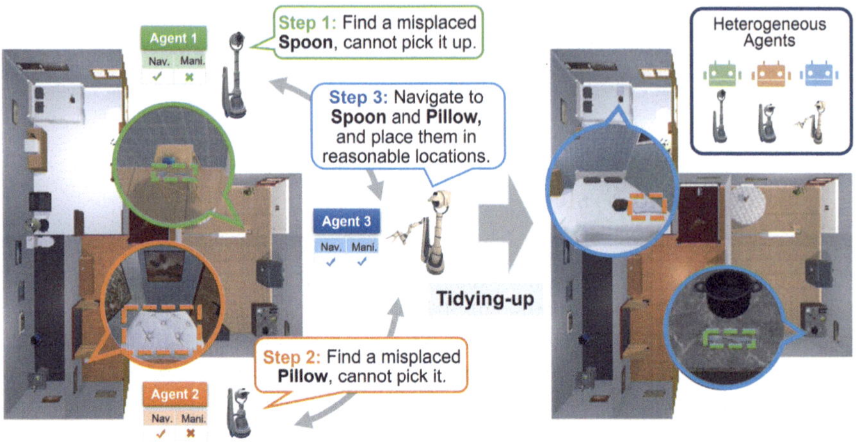

Fig. 3.11 An overview of the heterogeneous multi-agent room tidying-up task

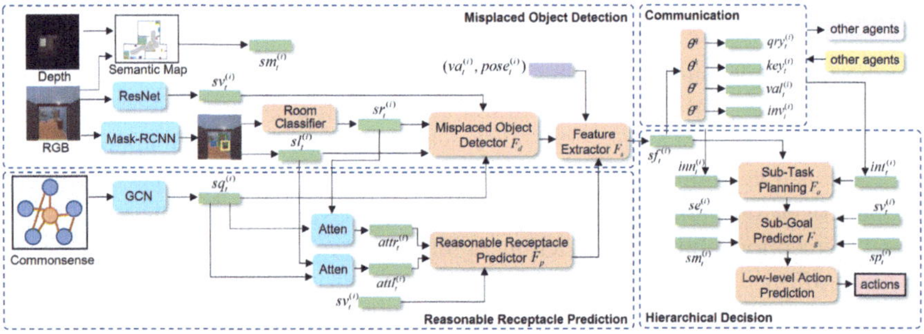

Fig. 3.12 An overview of the framework of the proposed heterogeneous multi-agent collaboration perception-action loop

perception observation $e_t^{(i)}$ and the communication information $c_t^{(i)}$ from others. Each agent $A^{(i)}$ needs to learn an action strategy $\pi^{(i)}$ to generate the action $a_t^{(i)}$ from its action space $I^{(i)}$, $a_t^{(i)} = \pi^{(i)}(e_t^{(i)}, c_t^{(i)})$. The objective of this tidying-up task is to learn the optimal collaboration action strategy $\pi^{(i)}$ for each agent such that they can collaboratively find all misplaced objects and place them in reasonable locations to make $D(o_j, p_j^*, r_j^*) = True, j = 1, 2, \cdots, M$ with as few steps as possible, where (o_j, p_j^*, r_j^*) denotes the location of objects after all agents stop. We demonstrate the heterogeneous multi-agent room tidying-up task in Fig. 3.11.

3.4.2 *Perception-Action Framework*

The proposed perception-action framework for the heterogeneous room tidying-up task consists of four main modules, which is shown in Fig. 3.12. The misplaced object detector judges whether there exists an object placed in an unreasonable location. The reasonable receptacle predictor generates a reasonable location to place the misplaced objects. The communication module transmits the information to other heterogeneous agents with the group communication mechanism. The hierarchical decision module predicts the next subtask, subgoal, and actions for each agent to execute.

3.4.2.1 Misplaced Object Detector

To obtain effective information of misplaced objects, the misplaced object detector is built to judge the rationality of object placement. Each agent builds a top-down semantic map with the input RGB and depth using a method similar to semantic mapping in [12]. At each subgoal step, the agent obtains a local semantic map of size $G \times G \times (K_{total} + 2)$, where G indicates the size of the local map, K_{total} is the number of object categories, and another two channels denote the obstacles and explored areas, respectively. Then agent $A^{(i)}$ generates the map embedding $sm_t^{(i)}$ with a pretrained scene encoder consisting of multiple convolutional layers similar to that in [13]. We extract spatial relationships among objects, receptacles, and room types that exist in our dataset from Visual Genome [11] to form a graph structure. We regard these relationships as the commonsense knowledge. We use the Glove embedding to encode the class label, concatenate with visual embeddings $sv_t^{(i)}$ extracted by ResNet [6] from visual observations as node features, and use GCN [9] to obtain the commonsense embedding $sq_t^{(i)}$. We pretrain Mask R-CNN [5] with around 50k images and ground truth generated from training houses in ProcTHOR-10K. It is used to detect objects and obtain the fused detection feature of all detected objects $sl_t^{(i)}$. We construct a room classifier including two linear layers to predict the room type where the agent is currently located based on the placement relationship of detected objects and extract room embedding $sr_t^{(i)}$. The misplaced object detector F_d, a classifier consisting of two linear layers, is utilized to fuse embeddings and detect whether the object is in a reasonable location. $det_t^{(i)} = F_d(sq_t^{(i)}, sv_t^{(i)}, sl_t^{(i)}, sr_t^{(i)})$. $det_t^{(i)} = 1$ denotes the object is misplaced, otherwise $det_t^{(i)} = 0$.

3.4.2.2 Reasonable Receptacle Predictor

The reasonable receptacle predictor is built to obtain effective information of the target receptacle and benefit the subsequent subtask planning. The detection feature $sl_t^{(i)}$ and the commonsense embedding $sq_t^{(i)}$ are fed into an attention layer to obtain the attention $attl_t^{(i)}$. The room embedding $sr_t^{(i)}$ and $sq_t^{(i)}$ are also fed into an attention layer to generate the attention $attr_t^{(i)}$. The reasonable receptacle predictor F_p including linear layers fuses the visual embedding $sv_t^{(i)}$ with the attention embedding and predicts the reasonable location including receptacle p_t and room type r_t for the misplaced object o_t, and then we have $(p_t, r_t) = F_p(sv_t^{(i)}, attl_t^{(i)}, attr_t^{(i)})$.

3.4.2.3 Communication

Before communication, each agent generates a characteristic vector $va^{(i)}$ based on their capability and property, $va^{(i)} = (Nav^{(i)}, Mani^{(i)}, Hei^{(i)})$, and obtains a pose vector $pose_t^{(i)} = (x_t^{(i)}, y_t^{(i)}, rot_t^{(i)})$. The state feature extractor F_s including linear layers and the LSTM [7] layer is utilized to obtain the fused state feature $sf_t^{(i)} = F_s(va^{(i)}, pose_t^{(i)}, fd_t^{(i)}, fp_t^{(i)})$, where $fd_t^{(i)}, fp_t^{(i)}$ are features extracted from the layer before the last classification layer of F_d and F_p. We propose the *Handshake-based Group Communication* to save the communication amounts and achieve a balance between good task performance and communication amounts. In this module, the agent generates the query vector $qry_t^{(i)}$, key vector $key_t^{(i)}$, value vector $val_t^{(i)}$, and inter-group information vector $inv_t^{(i)}$ based on its state feature $[qry_t^{(i)}, key_t^{(i)}, val_t^{(i)}, inv_t^{(i)}] = [\theta^q, \theta^k, \theta^v, \theta^e](sf_t^{(i)})$, where $\theta^q, \theta^k, \theta^v, \theta^e$ are the corresponding vector generators consisting of linear layers, respectively. At the t-th subgoal step, each agent sends $qry_t^{(i)}$ to others and calculates the scaled inner product attention of received query vectors and its key vector $attc_{ij} = \frac{qry_t^{(j)} key_t^{(i)}}{\sqrt{d}}$, where d is the dimension of $qry_t^{(j)}$ and $key_t^{(i)}$. Then a communication matrix $T_t = \sigma([attc_{ij}]_{N\times N})$ is obtained, where $attc_{ij}$ indicates the effectiveness of the information send from $A^{(j)}$ to $A^{(i)}$, and σ denotes row-wise softmax function. When communicating, we set two thresholds δ and μ. If $attc_{ii} < \delta$, $A^{(i)}$ needs to receive messages from other agents, since its own information is not efficient. If $attc_{ij} > \mu(i \neq j)$, $A^{(i)}$ can receive messages from $A^{(j)}$. Based on T_t, agents can be implicitly divided into several groups. Intra-group communication exchanges the state information from agents in the same group to complete short-term subtasks, and $A^{(i)}$ obtain the intra-group communication information $inn_t^{(i)} = \sum_{(j\neq i, att_{ij}>\mu)} att_{ij} \cdot val_t^{(j)}$. Inter-group communication conveys higher-level information on the task allocation across different groups, which is beneficial for agents to make future decisions in the long-term task completion. $A^{(i)}$ obtains the inter-group communication information $int_t^{(i)} = inv_t^{(j)}, j = argmax(avg(att_{*j}))$, where avg is the average operation and $*$ denotes the ids of agents that belong to the same group as $A^{(i)}$. $inn_t^{(i)}$ and $int_t^{(i)}$ are utilized to generate the subtask. An example is shown in Fig. 3.13.

In the training process, in order to ensure the generation of communication information is differentiable, it is assumed agents can obtain state features of all other agents. At each subgoal step, $A^{(i)}$ obtains the intra-group communication information $inn_t^{(i)} = \sum_{j=1}^{N} att_{ij} \cdot val_t^{(j)}$ and inter-group information $int_t^{(i)} = \sum_{j=1}^{N}(1 - att_{ij}) \cdot inv_t^{(j)}$. In the inference process, agents only obtain their own

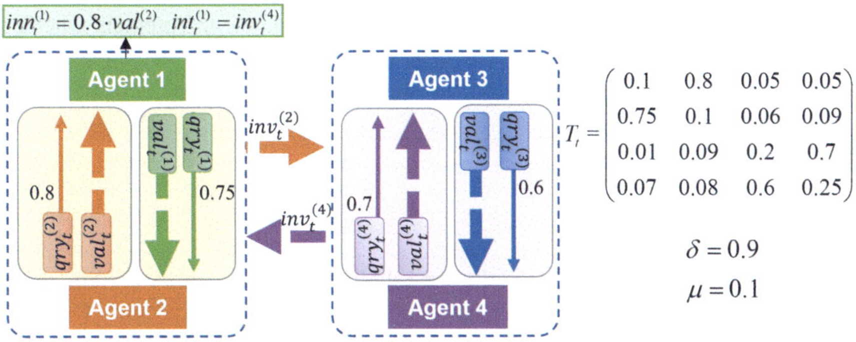

Fig. 3.13 Details of the communication process. Based on T_t, at that step, *Agent 1* and *Agent 2* would form a communication group, and *Agent 3* and *Agent 4* form a group. The intra-group and inter-group communication information of $A^{(1)}$ is obtained based on T_t, δ, and μ

state features and complete the communication process according to latent divided communication groups under the threshold δ, μ as mentioned above. We set $\mu = 0.2$, $\delta = 0.8$ in our experiments.

3.4.2.4 Hierarchical Decision

The hierarchical decision module is designed to predict the next subtask, subgoal, and low-level actions in turn according to the decision level, which avoids generating invalid or illogical action sequences.

Subtask Planning The subtask planning part generates the subtask for each agent to complete. In this task, two types of subtasks can be executed: *Explore* and *Place*. In *Explore*, the agent explores the scene and searches for misplaced objects. The subtask *Place* contains three parameters, denoted as $(Place, o_t, p_t, r_t)$, which means to put the misplaced object o_t to the receptacle p_t in the room type r_t. A subtask predictor F_a, which consists of the LSTM layer and linear layers, is utilized to predict the type and parameters of the executed subtask with $sf_i^{(i)}$, $inn_i^{(i)}$, and $int_i^{(i)}$. We use the output of the layer before the last prediction layer to obtain the embedding $se_t^{(i)}$ of the predicted subtask $se_t^{(i)} = F_a(sf_t^{(i)}, inn_t^{(i)}, int_t^{(i)})$.

Action Decision Each agent encodes $pose_t^{(i)}$ to get the pose embedding $sp_t^{(i)}$. A subgoal predictor F_g, which consists of the LSTM layer and linear layers, is utilized to predict the next subgoal $sg_t^{(i)}$ with $se_t^{(i)}$, $sm_t^{(i)}$, $sp_t^{(i)}$, and $sv_t^{(i)}$. $sg_t^{(i)} = (\Delta x_t^{(i)}, \Delta y_t^{(i)}, \Delta rot_t^{(i)}, ope_t^{(i)}, stop_t^{(i)}) = F_g(se_t^{(i)}, sm_t^{(i)}, sp_t^{(i)}, sv_t^{(i)})$, where $\Delta x_t^{(i)}$, $\Delta y_t^{(i)}$ are the number of movement grids with 0.25 meters of the subgoal within 1 meter in the x and y axis, respectively, $\Delta rot_t^{(i)}$ is the rotation angle (0, 90, 180, or 270), and $ope_t^{(i)}$ denotes whether performing action *PickUp*,

PutDown, *Drop*, or *NoAction* after reaching the subgoal. $stop_t^{(i)}$ predicts whether the agent stops or not after this subgoal step. Then, a low-level action strategy based on the shortest path algorithm is utilized to generate specific actions from $I^{(i)}$ to reach the subgoal.

3.4.2.5 Model Training

Since the functions of each module are relatively independent, we train them separately. We first train the room classifier with supervised learning. The loss function is the cross entropy loss between the predicted and the true type of the room where the agent is located. Then, we train F_d and F_p with supervised learning. We generate true labels of training samples based on the metadata of the scene and train them with the parameters of the room classifier fixed. The loss function is defined as

$$Loss_{det\_rec} = \alpha L_{mis} + \beta L_{rec1} + \lambda L_{room1}, \tag{3.7}$$

where L_{mis} indicates the binary cross entropy of whether the object is misplaced. L_{rec1}, L_{room1} denotes the multi-label classification loss between the predicted and true reasonable types of the receptacle and the room. α, β, λ are the hyper-parameters. In our implementation, we set α, β, λ to 1.

We use behavior cloning with generated expert demonstrations to train the subtask planning part and the action decision part in the hierarchical decision module. The loss function is defined as follows:

$$\begin{aligned} Loss_{subtask} &= L_{task} + \gamma_1 L_{obj} + \delta_1 L_{rec2} + \theta_1 L_{room2} \\ Loss_{subgoal} &= L_{x_{loc}} + L_{y_{loc}} + \gamma_2 L_{rot} + \delta_2 L_{ope} + \theta_2 L_{stop}, \end{aligned} \tag{3.8}$$

where L_{task}, L_{obj}, L_{rec2}, L_{room2} denotes the cross entropy loss between the predicted subtask type, $o_t^{(i)}$, $p_t^{(i)}$, $r_t^{(i)}$, and those in demonstrations, respectively. $L_{x_{loc}}$, $L_{y_{loc}}$, L_{rot}, L_{ope}, L_{stop} denotes the cross entropy loss between the predicted results and those in demonstrations, respectively. We set hyper-parameters γ_1, γ_2, δ_1, δ_2, θ_1, θ_2 to be 1.

3.4.3 *Simulation Experiments*

3.4.3.1 Dataset

To evaluate the proposed heterogeneous multi-agent perception-action framework, we create the novel dataset for the heterogeneous multi-agent tidying-up tasks based on the ProcTHOR-10K [4]. In this task, the observations provided to each agent are

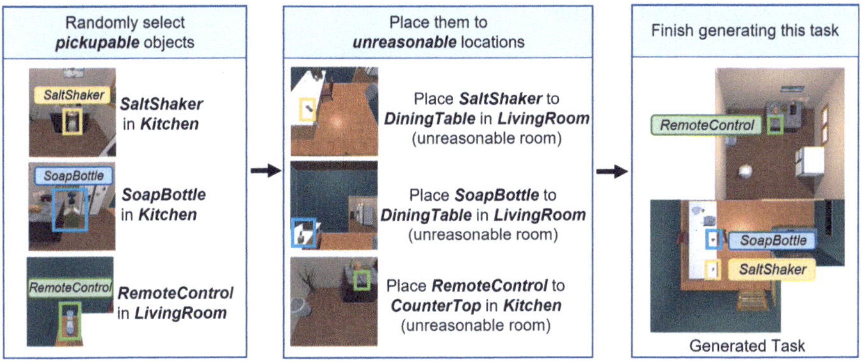

Fig. 3.14 The process of generating the room tidying-up task from the original scene. For each task, we randomly select k objects from the house and then change their current locations to unreasonable receptacles or unreasonable rooms

the first-view RGB-D images in the visual modality and the pose information of the agent's global 3D coordinate positions. In each task, several objects are placed in unreasonable locations, and multiple heterogeneous agents are required to find these misplaced objects and put them to the reasonable locations. ProcTHOR-10K is a generated dataset of houses with multiple rooms built based on AI2-THOR [10], including 10,000 training houses, 1000 validation houses, and 1000 testing houses, and each room belongs to one type of *Kitchen*, *LivingRoom*, *Bedroom*, and *Bathroom*. There are 118 possible object categories in the whole dataset, and 43 of them are receptacles. We propose a method to generate multi-agent tidying-up tasks in ProcTHOR-10K. Since every single room in ProcTHOR-10K is built based on AI2-THOR, we regard the object placement relationships in AI2-THOR as the reasonable placement constraints in ProcTHOR-10K. For each object in O_{pick}, we generate the triples of reasonable candidate locations $Re_i = \{(o_i, p_j, r_j)\}$ from the metadata obtained from ProcTHOR-10K, where o_i can be put on receptacle p_j in room r_j. We generate 10 meta-tasks for each house, and the process is shown in Fig. 3.14. For each meta-task, we randomly select k objects to change their current location $(o_i, p_i^{cur}, r_i^{cur})$ to an unreasonable location $(o_i, p_i^{new}, r_i^{new})$, and k is randomly chosen from $\{1, 2, 3, 4, 5\}$. We place the misplaced object to an unreasonable receptacle or drop the object to a random location in the house. All meta-tasks are divided into *Single-Room* (*Single*) and *Cross-Room* (*Cross*) tasks. In *Single* tasks, all k misplaced objects satisfy $r_j^{new} = r_j^{cur}$, and in *Cross* tasks, at least one misplaced object satisfies $r_j^{new} \neq r_j^{cur}$. For each meta-task, we randomly generate five initial positions for multiple heterogeneous agents. We also generate expert demonstration using heuristic shortest path methods including the subtask planning and subgoal planning demonstration for imitation learning based on the metadata and scene information of each task.

3.4.3.2 Experiment Setting

We focus on heterogeneous multi-agent collaboration, and we evaluate the proposed collaboration strategy in the room tidying-up task. To effectively and efficiently train and verify our model, we randomly select 120 scenes from the newly built dataset, 80 houses for training, 20 for validation, and 20 for testing. The number of scenes in our experiment is the same as that in other embodied tasks in AI2-THOR, which is widely used in different embodied tasks. Each house contains 10 different meta-tasks, and each meta-tasks contains five different initial positions for heterogeneous agents. The time for each agent to execute each step is the same, and the agent can obtain messages without information loss and delay during communication. Each agent only has the partial observation and does not have the panoramic view. Before executing each low-level action, each agent judges whether it will collide with others and, if so, replans the low-level action sequences.

We consider two different heterogeneous settings. In *Setting I*, there are three agents with the same visual perception ability but different action abilities and morphologies. Specifically, $A^{(1)}$ only has the navigation ability with the low height, $Nav^{(1)} = 1, Mani^{(1)} = 0, Hei^{(1)} = 0$. $A^{(2)}$ only has the navigation ability and its height is high, $Nav^{(2)} = 1, Mani^{(2)} = 0, Hei^{(2)} = 1$. $A^{(3)}$ has both the navigation and manipulation abilities, and its height is high, $Nav^{(3)} = 1, Mani^{(3)} = 1, Hei^{(3)} = 1$. In *Setting II*, there are four heterogeneous agents. $A^{(1)}$, $A^{(2)}$, and $A^{(3)}$ are the same as *Setting I*. $A^{(4)}$ has the same abilities and the same morphology with $A^{(3)}$. The reasons we select the *Setting I* and *Setting II* are that these two settings can represent the general heterogeneous collaboration scenarios, and corresponding experimental results can demonstrate the generalization of our model across different numbers and heterogeneous settings of agents. For each setting, there are 4000 training tasks (80×10×5), 1000 validation tasks (20×10×5), and 1000 testing tasks. Training tasks contain 2260 *Single* and 1740 *Cross* tasks. Validation tasks contain 550 *Single* and 450 *Cross* tasks. Testing tasks contain 565 *Single* and 435 *Cross* tasks.

3.4.3.3 Evaluation Metrics

We propose six metrics to evaluate the proposed model:

1. **Success rate (SR)**: $SR = \frac{1}{N_{epi}} \sum_{i=1}^{N_{epi}} R_i$, where $R_i = 1$ if all misplaced objects are replaced in reasonable locations, otherwise $R_i = 0$, and N_{epi} is the number of tasks. It denotes the strict success rate.
2. **%PartialSuccess (%PS)**: $\%PS = \frac{1}{N_{epi}} \sum_{i=1}^{N_{epi}} \frac{K_i^{suc}}{K_i}$, where K_i is the number of misplaced objects in the i-th task, and K_i^{suc} is the number of misplaced objects that are replaced to reasonable locations. It reflects the proportion of misplaced objects that are successfully tidied.

3. **%FindMisObj (%FM)**: $\%FM = \frac{1}{N_{epi}} \sum_{i=1}^{N_{epi}} \frac{K_i^{det}}{K_i}$, where K_i^{det} is the number of misplaced objects that are detected and picked up by agents in this task. It can reflect the effect of misplaced objects detector to some extent.
4. **#PathLength (#PL)**: $\#PL = \frac{1}{N_{epi}} \sum_{i=1}^{N_{epi}} Len_i$, where Len_i is the number of steps in the multi-agent trajectory when synchronously completing the i-th task. It evaluates the efficiency of completing the task. The fewer steps, the higher the efficiency of the multi-agent collaboration.
5. **AvgCom (ACm)**: $ACm = \frac{1}{N_{epi}} \sum_{i=1}^{N_{epi}} \frac{Total_i}{Len_i \cdot N}$, where $Total_i$ is the total dimensions of communication messages among agents. It evaluates the average communication amounts per agent per step.
6. **CommEffi(CE)**: $CE = \max(0, \frac{10000 \cdot (Suc - SucSA)}{ACm})$, where $SucSA$ is the success rate of the single-agent model. It evaluates the proportion of efficiency improvement of different communication methods. The higher CES, the higher the efficiency of the communication mechanism.

3.4.3.4 Quantitative Analysis

We evaluate the proposed model using different action decision strategies and communication methods in testing tasks. For different action strategies, we compare with *SA*, *SA(Ora)*, *Random*, and *QMIX* model. *SA* only utilizes one agent with both navigation and manipulation abilities to complete the task. In *SA(Ora)*, the objects' locations are known to the single agent in *SA*. *Random* randomly generates actions for each agent. *QMIX* is a MARL method based on [15], in which agents fuse their state features with others and directly predict the low-level actions. For different communication methods, we consider both decentralized and centralized methods. In *CondCom*, only when an agent detects the misplaced object, it broadcasts its state and maps to others. *CmprCom* utilizes an encoder-decoder structure to compress the communication messages and decode received messages from other agents [12]. In *IntenCom*, agents exchange state information, infer other agent's next subgoal, and decide their next subtask [20]. *CentralCom* adopts a central node to obtain the messages from all agents. In *BroadCom*, each agent broadcasts its messages to all agents and determines actions with received messages. All these communication baselines are demonstrated in Fig. 3.15.

We show quantitative results of *Setting I* and *Setting II* in Tables 3.9 and 3.10, respectively. The maximum number of steps per task is 300. The gray background indicates the methods are regarded as the upper bound reference that shares full information during communication. The best performances except for central and broadcast methods are bolded. The results in *Setting I* and *Setting II* indicate that our model can generalize to different numbers and heterogeneous settings of agents. Meanwhile, the heterogeneous multi-agent tidying-up task is challenging. $\%FM$ is relatively high, but the success rate is relatively low, which indicates that moving

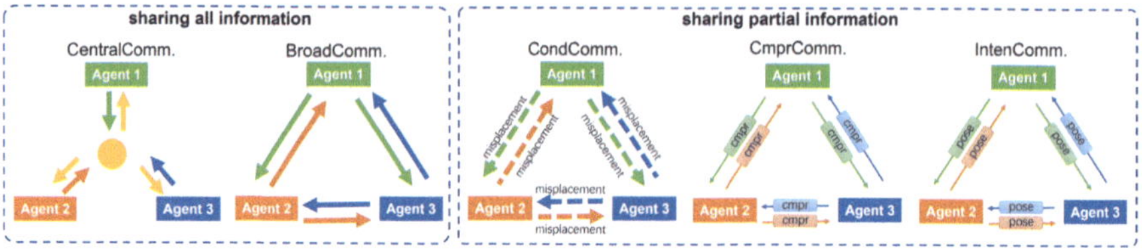

Fig. 3.15 Communication baselines in experiments. The thickness of lines represents the communication amount. Solid lines indicate the information is passed at each step, and dotted lines indicate the information is passed only when certain conditions are met. Rectangles on lines represent delivered vectors. The rectangle on the line represents the delivered specific vectors

misplaced objects to a reasonable location is challenging, since agents may not reason the proper receptacle or not generate valid actions. The success rate of *SA* and *SA(Ora)* is lower, which demonstrates that collaboration among heterogeneous multi-agents can improve the accuracy and efficiency. Performance of *QMIX* is relatively low, indicating that this MARL method is not effective enough to deal with visual information, and it is difficult to directly predict low-level actions. As for different communication methods, since *CentralCom* utilizes a central node to process all information, its task completion performance (SR,$\%PS$,$\%FM$,$\#PL$) can be regarded as the upper bound of multi-agent models. Performances of *Ours*, *CondCom*, *CmprCom*, and *IntenCom* in task completion are lower than *CentralCom* and *BroadCom*, because agents share partial information with others for decision-making. Our model performs better than other methods that share partial messages in task completion. On the other hand, ACm of *CentralCom* and *BroadCom* are the largest, and with the increased number of agents, their ACm increases greatly but declines in CE are relatively larger, which indicates that they are not cost-effective enough. Although *IntenCom* uses the smallest ACm, its CE is not high, since its Suc is not good. Our method can save communication amounts, achieve good performance, and obtain the highest communication efficiency.

3.4.4 Qualitative Analysis

We demonstrate successful samples with four agents in *Setting II* in Fig. 3.16, which demonstrates that the proposed model can make effective use of different abilities of heterogeneous to improve efficiency. We also show two failed cases in Fig. 3.17, where agents place the misplaced object in an unreasonable location or fail to detect the misplaced objects.

Table 3.9 Quantitative results in heterogeneous multi-agent room tidying-up task in *Setting I*

Methods	$SR(\uparrow)$			$\%PS(\uparrow)$			$\%FM(\uparrow)$			$\#PL(\downarrow)$			$ACm(\downarrow)$			$CE(\uparrow)$		
	Sin	Cro	All	Sin	Cro	All	Sin	Cro	All	Sin	Cro	All	Sin	Cro	All	Sin	Cro	All
SA	4.5	3.0	3.8	9.6	8.1	8.9	52.5	48.3	50.7	278.6	294.5	285.5	–	–	–	–	–	–
SA(Ora)	9.0	6.8	8.0	18.1	16.3	17.3	70.3	68.1	69.3	184.8	197.1	190.1	–	–	–	–	–	–
Random	0	0	0	0	0	0	0.3	0	0.2	300.0	300.0	300.0	0	0	0	–	–	–
QMIX	3.2	1.5	2.5	8.2	6.9	7.6	47.2	42.1	45.0	293.1	300.0	296.1	512.0	512.0	512.0	0	0	0
CondCom	8.1	6.6	7.5	17.6	15.1	16.5	71.2	70.9	71.0	192.3	263.1	223.1	523.1	501.3	513.6	0.7	0.7	0.7
CmprCom	6.7	5.3	6.1	16.7	13.1	15.1	69.5	67.7	68.7	188.3	253.3	216.6	220.0	220.0	220.0	1.0	1.0	1.0
IntenCom	4.7	3.2	4.0	13.6	10.1	12.1	64.5	60.9	62.9	214.6	267.5	237.6	**20.0**	**20.0**	**20.0**	1.0	1.0	1.0
Ours	**12.1**	**7.9**	**10.3**	**19.4**	**16.0**	**17.9**	**73.9**	**72.6**	**73.3**	**186.1**	**252.6**	**215.0**	363.1	366.5	364.6	**2.1**	**1.4**	**1.8**
BroadCom	13.0	8.9	11.2	20.1	18.4	19.4	74.2	73.1	73.7	183.3	245.9	210.5	820.0	820.0	820.0	1.0	0.7	0.9
CentralCom	13.1	9.2	11.4	20.5	16.6	18.8	75.5	73.9	74.8	183.2	243.1	209.3	820.0	820.0	820.0	1.0	0.8	0.9

[1] All values of SR, $\%PS$ and $\%FM$, are percentages in this table.
[2] *Sin* and *Cro* in this table represent the experimental results in *Single-Room* and *Cross-Room* tasks, respectively.

Table 3.10 Quantitative results in heterogeneous multi-agent room tidying-up task in *Setting II*

Methods	$SR(\uparrow)$			$\%PS(\uparrow)$			$\%FM(\uparrow)$			$\#PL(\downarrow)$			$ACm(\downarrow)$			$CE(\uparrow)$		
	Sin	Cro	All	Sin	Cro	All	Sin	Cro	All	Sin	Cro	All	Sin	Cro	All	Sin	Cro	All
SA	4.5	3.0	3.8	9.6	8.1	8.9	52.5	48.3	50.7	278.6	294.5	285.5	–	–	–	–	–	–
SA(Ora)	9.0	6.8	8.0	18.1	16.3	17.3	70.3	68.1	69.3	184.8	197.1	190.1	–	–	–	–	–	–
Random	0	0	0	0	0	0	0.5	0	0.3	300.0	300.0	300.0	0	0	0	–	–	–
QMIX	3.5	2.1	2.9	9.1	7.6	8.4	48.3	42.5	45.8	290.5	296.3	292.9	512.0	512.0	512.0	0	0	0
CondCom	9.0	7.2	8.2	17.9	15.6	16.9	72.0	71.1	71.6	188.5	230.7	206.9	620.6	613.1	617.3	0.7	0.7	0.7
CmprCom	7.3	5.9	6.7	17.3	13.9	15.8	70.3	68.3	69.4	183.1	226.9	202.1	330.0	330.0	330.0	0.8	0.9	0.8
IntenCom	4.8	3.3	4.1	13.9	11.2	12.7	66.1	62.0	64.3	210.1	255.3	229.8	**30.0**	**30.0**	**30.0**	1.0	1.0	1.0
Ours	**12.9**	**9.0**	**11.2**	**20.3**	**16.9**	**18.8**	**74.5**	**73.9**	**74.3**	**166.9**	**221.3**	**190.6**	472.5	478.3	475.0	**1.8**	**1.3**	**1.6**
BroadCom	13.6	9.4	11.8	20.9	18.9	20.0	74.8	74.1	74.5	161.1	213.3	183.8	1230.0	1230.0	1230.0	0.7	0.5	0.6
CentralCom	13.8	9.7	12.0	21.1	17.3	19.4	76.0	74.5	75.3	160.6	213.1	183.4	1230.0	1230.0	1230.0	0.8	0.5	0.7

[1] All values of SR, $\%PS$ and $\%FM$, are percentages in this table.
[2] *Sin* and *Cro* in this table represent the experimental results in *Single-Room* and *Cross-Room* tasks, respectively.

Fig. 3.16 The successful samples in *Setting II*. In the top-down view, the color of bounding boxes in the left indicates which agent detects the specific misplaced objects, and the color of arrows indicates which agent puts the misplaced objects to reasonable locations

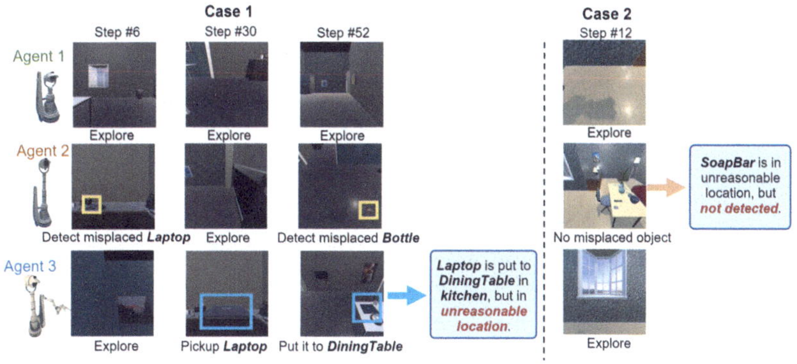

Fig. 3.17 Failed cases in *Setting I*. In *Case 1*, *Agent 3* places *Laptop* to an unreasonable location. In *Case 2*, *Agent 2* fails to detect misplaced *SoapBar*

3.4.5 Ablation Experiments

The results of ablation experiments in *Setting I* and *Setting II* are shown in Tables 3.11 and 3.12, respectively. *w/o Kno* removes the commonsense knowledge, and its performance drops obviously, indicating that the commonsense knowledge is important for scene reasoning in this task. *w/o MisObjDec* and *w/o ReaRecPre* remove the misplaced object detector and the reasonable receptacle predictor, respectively. Since these two modules can provide a clear target object and a clear receptacle to put the object for subsequent subtasks generating, their performance decreases. *w/o HierDec* directly generates low-level actions with the state features. Because it is difficult to directly learn the relations between state features and low-

Table 3.11 Ablation experiments results in *Setting I*

	$SR(\uparrow)$			$\%PS(\uparrow)$			$\%FM(\uparrow)$			$\#PL(\downarrow)$			$ACm(\downarrow)$			$CE(\uparrow)$		
Methods	Sin	Cro	All	Sin	Cro	All	Sin	Cro	All	Sin	Cro	All	Sin	Cro	All	Sin	Cro	All
w/o Kno	4.2	3.0	3.7	8.7	8.0	8.4	42.5	36.6	39.9	235.1	267.3	249.1	363.5	366.6	364.8	0	0	0
w/o MisObjDec	5.1	4.6	4.9	10.9	9.1	10.1	51.5	47.1	49.6	234.6	264.5	247.6	363.3	366.6	364.7	0.2	0.4	0.3
w/o ReaRecPre	6.3	5.8	6.1	12.0	9.8	11.0	55.1	50.9	53.3	231.1	260.3	243.8	363.3	366.9	364.9	0.5	0.8	0.6
w/o Com	4.5	3.0	3.8	9.6	8.1	8.9	52.5	48.3	50.7	278.6	294.5	285.5	0	0	0	–	–	–
w/o HierDec	5.5	4.9	5.2	10.6	9.3	10.0	57.3	53.2	55.5	266.3	284.5	274.2	366.1	367.2	366.6	0.3	0.5	0.4
Ours	12.1	7.9	10.3	19.4	16.0	17.9	73.9	72.6	73.3	186.1	252.6	215.0	363.1	366.5	364.6	2.1	1.4	1.8
Ora MisObjDec	13.2	8.6	11.2	20.5	17.2	19.1	92.3	90.7	91.6	183.6	249.9	212.4	362.9	366.2	363.6	2.4	1.5	2.0
Ora ReaRecPre	17.6	12.1	15.2	23.1	21.0	22.2	73.9	72.6	73.3	179.3	245.1	207.9	363.2	366.3	364.5	3.6	3.2	3.4

[1] All values of SR, $\%PS$ and $\%FM$, are percentages in this table.
[2] *Sin* and *Cro* in this table represent the experimental results in *Single-Room* and *Cross-Room* tasks, respectively.

Table 3.12 Ablation experiments results in *Setting II*

	$SR(\uparrow)$			$\%PS(\uparrow)$			$\%FM(\uparrow)$			$\#PL(\downarrow)$			$ACm(\downarrow)$			$CE(\uparrow)$		
Methods	Sin	Cro	All	Sin	Cro	All	Sin	Cro	All	Sin	Cro	All	Sin	Cro	All	Sin	Cro	All
w/o Kno	4.9	3.3	4.2	10.3	9.6	1.0	45.2	37.5	41.9	227.0	258.5	240.7	473.3	479.2	475.9	0.09	0.06	0.08
w/o MisObjDec	5.8	5.2	5.5	11.8	10.3	11.1	52.3	48.0	50.4	225.5	256.1	238.8	473.1	479.0	475.7	0.3	0.5	0.4
w/o ReaRecPre	7.5	6.9	7.2	12.8	10.5	11.8	56.0	51.5	54.0	220.6	251.9	234.2	473.3	478.9	475.8	0.6	0.8	0.7
w/o Com	4.9	3.3	4.2	10.5	9.1	9.9	53.5	49.6	51.8	273.1	280.6	276.4	0	0	0	–	–	–
w/o HierDec	6.3	5.5	6.0	11.1	9.9	10.6	58.1	53.7	56.2	260.1	273.6	266.2	475.3	478.8	477.3	0.4	0.5	0.4
Ours	12.9	9.0	11.2	20.3	16.9	18.8	74.5	73.9	74.3	166.9	221.3	190.6	472.5	478.3	475.0	1.8	1.3	1.6
Ora.MisObjDec	13.9	9.8	12.1	21.2	17.6	19.6	92.7	91.7	92.2	163.8	219.1	187.9	472.3	478.5	475.0	2.0	2.0	2.0
Ora ReaRecPre	17.8	14.0	16.1	24.7	21.9	23.5	73.9	72.6	73.3	160.2	215.7	184.3	472.3	467.3	470.1	2.8	2.9	2.8

[1] All values of SR, $\%PS$ and $\%FM$, are percentages in this table.
[2] *Sin* and *Cro* in this table represent the experimental results in *Single-Room* and *Cross-Room* tasks, respectively.

level actions, this model would generate some invalid actions, leading to worse performance. Furthermore, we also consider *Ora MisObjDec* and *Ora ReaRecPre*, which utilize the ground truth of misplaced object detector and reasonable receptacle predictor in our model, respectively. $\%FM$ of *Ora MisObjDec* illustrates that although the misplaced object detection is accurate, it is not easy to navigate to the misplaced object and pick it up. The performance of *Ora ReaRecPre* shows that the accuracy of reasonable receptacle prediction is important in task completion, and it is also challenging to find all misplaced objects as well as predict the subtask and subgoal although the reasonable receptacle prediction is accurate.

3.5 Summary

In this chapter, we introduce the structure of the centralized perception-action loop in the homogeneous embodied multi-agent system in the multi-agent task assignment problem. We analyze the decentralized perception-action loop in the homogeneous multi-agent system in the collaborative object search task. We demonstrate the perception-action loop in the heterogeneous embodied multi-agent system in the room tidying-up task and compare the performance of different communication mechanisms. We also build datasets to evaluate the corresponding structures of perception-action loops in different settings and demonstrate the experiments in the simulation environment.

References

1. Anderson P, Chang A, Chaplot DS, Dosovitskiy A, Gupta S, Koltun V, Kosecka J, Malik J, Mottaghi R, Savva M, Zamir AR (2018) On evaluation of embodied navigation agents. arXiv preprint arXiv:1807.06757
2. Bundy A, Wallen L (1984) Breadth-first search. In: Catalogue of artificial intelligence tools, pp 13–13. Springer, Berlin
3. Deitke M, Han W, Herrasti A, Kembhavi A, Kolve E, Mottaghi R, Salvador J, Schwenk D, VanderBilt E, Wallingford M, Weihs L, Yatskar M, Farhadi A (2020) RoboTHOR: An Open Simulation-to-Real Embodied AI Platform. In: CVPR
4. Deitke M, VanderBilt E, Herrasti A, Weihs L, Salvador J, Ehsani K, Han W, Kolve E, Farhadi A, Kembhavi A, Mottaghi R (2022) ProcTHOR: Large-Scale Embodied AI Using Procedural Generation. In: NeurIPS
5. He K, Gkioxari G, Dollár P, Girshick R (2017) Mask R-CNN. In: Proceedings of the IEEE international conference on computer vision, pp 2961–2969
6. He K, Zhang X, Ren S, Sun J (2016) Deep residual learning for image recognition. In: Proceedings of the IEEE conference on computer vision and pattern recognition, pp 770–778
7. Hochreiter S, Schmidhuber J (1997) Long short-term memory. Neural Comput 9(8):1735–1780
8. Jain U, Weihs L, Kolve E, Farhadi A, Lazebnik S, Kembhavi A, Schwing A (2020) A cordial sync: going beyond marginal policies for multi-agent embodied tasks. In: ECCV, pp 471–490. Springer, Berlin
9. Kipf TN, Welling M (2017) Semi-supervised classification with graph convolutional networks. In: ICLR

10. Kolve E, Mottaghi R, Han W, VanderBilt E, Weihs L, Herrasti A, Gordon D, Zhu Y, Gupta A, Farhadi A (2017) AI2-THOR: an interactive 3D environment for visual AI. arXiv preprint arXiv:1712.05474
11. Krishna R, Zhu Y, Groth O, Johnson J, Hata K, Kravitz J, Chen S, Kalantidis Y, Li LJ, Shamma DA, et al (2017) Visual genome: connecting language and vision using crowdsourced dense image annotations. Int J Comput Vis 123(1):32–73
12. Liu X, Guo D, Liu H, Sun F (2022) Multi-agent embodied visual semantic navigation with scene prior knowledge. IEEE Rob Autom Lett 7(2):3154–3161
13. Liu X, Li X, Guo D, Tan S, Liu H, Sun F (2022) Embodied multi-agent task planning from ambiguous instruction. In: RSS
14. Mnih V, Badia AP, Mirza M, Graves A, Lillicrap T, Harley T, Silver D, Kavukcuoglu K (2016) Asynchronous methods for deep reinforcement learning. In: International conference on machine learning, pp 1928–1937. PMLR, New York
15. Rashid T, Samvelyan M, De Witt CS, Farquhar G, Foerster J, Whiteson S (2020) Monotonic value function factorisation for deep multi-agent reinforcement learning. J Mach Learn Res 21(1):7234–7284
16. Schulman J, Wolski F, Dhariwal P, Radford A, Klimov O (2017) Proximal policy optimization algorithms. arXiv preprint arXiv:1707.06347
17. Shridhar M, Thomason J, Gordon D, Bisk Y, Han W, Mottaghi R, Zettlemoyer L, Fox D (2020) ALFRED: a benchmark for interpreting grounded instructions for everyday tasks. In: Proceedings of the IEEE/CVF conference on computer vision and pattern recognition, pp 10740–10749
18. Tan S, Sima K, Wang D, Ge M, Guo D, Liu H (2024) Self-supervised 3-d semantic representation learning for vision-and-language navigation. In: IEEE Transactions on Neural Networks and Learning Systems
19. Wang H, Wang W, Zhu X, Dai J, Wang L (2021) Collaborative visual navigation. arXiv preprint arXiv:2107.01151
20. Wang Y, Zhong F, Xu J, Wang Y (2021) ToM2C: target-oriented multi-agent communication and cooperation with theory of mind. arXiv:2111.09189
21. Yamauchi B (1998) Frontier-based exploration using multiple robots. In: Proceedings of the second international conference on Autonomous agents, pp 47–53
22. Yang W, Wang X, Farhadi A, Gupta A, Mottaghi R (2018) Visual semantic navigation using scene priors. arXiv preprint arXiv:1810.06543

Chapter 4
Embodied Cooperation in Multi-Agent System

Abstract In this chapter, we present a distributed, hierarchical locomotion control strategy designed for whole-body cooperation and capable of scaling to large groups of agents. Our approach leverages a hierarchical structure to decompose complex tasks into smaller, more manageable subtasks. By incorporating spatiotemporal continuity features, we establish the sequential logic essential for causal inference and cooperative behavior in tasks that unfold over time. This structure enables efficient and coordinated control strategies. Through training within this framework, we demonstrate enhanced adaptability and cooperation, resulting in superior task completion performance compared to traditional methods. Additionally, we introduce a set of environments as a benchmark for evaluating embodied cooperation.

4.1 Introduction

Cooperative execution of embodied tasks with multiple robots has long been a challenging research area. Recent studies have primarily focused on embodied manipulation cooperation among robotic arms or formation control within groups of mobile robots [12, 13]. However, multi-agent cooperation using whole-body and end-to-end locomotion control remains underexplored.

Some earlier works have demonstrated manipulation via locomotion [25], but these studies were limited to two-agent systems, and their scalability to larger populations of agents remains uncertain. In this chapter, we aim to advance multi-agent embodied cooperation by developing a distributed hierarchical locomotion control system that decomposes complex, coupled behaviors while preserving scalability for large swarm sizes. To facilitate implementation and validation, we introduce three benchmark scenarios in IsaacSim [23] as foundational environments for studying embodied cooperation (Fig. 4.1).

Training robots for specific tasks is often effectively achieved through reinforcement learning (RL), such as learning movement patterns [15], interactive behaviors [14], and logical inference in games [42]. While RL is known for its powerful exploration capabilities, and significant progress has been made in improving sampling efficiency [23], mastering a sequence of complex tasks through

H. Liu et al., *Embodied Multi-Agent Systems*, Machine Learning: Foundations, Methodologies, and Applications, https://doi.org/10.1007/978-981-96-5871-8_4

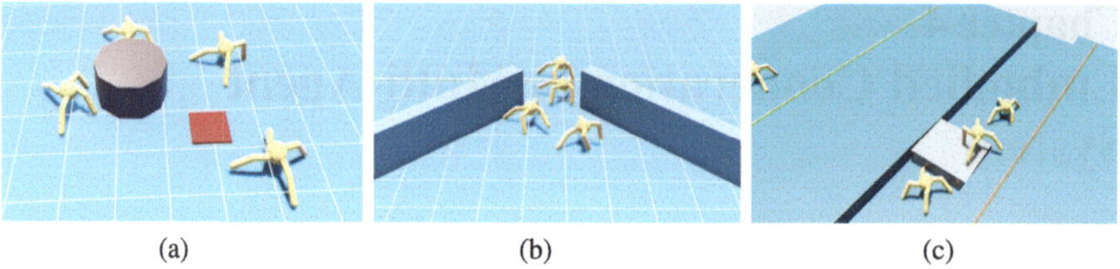

Fig. 4.1 Embodied cooperation environments. (Reproduced with permission from [11]). (a) Cooperative transport. (b) Corridor crossing. (c) Ravine bridging

pure search remains a significant challenge. Hierarchical reinforcement learning (HRL) partially addresses this issue by breaking down tasks into segments, helping to understand the logical relationships among "control, action, behavior, dynamic outcomes, and feedback." An increasing number of studies are adopting hierarchical approaches to decompose complex tasks [3, 4, 8, 16, 20, 33–35]. Building on this, we propose a fully decentralized HRL framework that naturally decomposes perception feature extraction, cooperative behaviors, and fundamental actions.

In previous work [25], embodied cooperation was achieved through locomotion control with a small number of robots. However, as the number of agents increases, these methods fail to scale effectively, which is a critical challenge in the multi-agent reinforcement learning (MARL) field [2, 24, 28, 37]. To address this issue, we propose a fully decentralized framework that enhances scalability and promotes cooperative behavior across larger swarms of intelligent agents. Additionally, we construct a set of scenarios for whole-body embodied cooperation in IsaacSim (Fig. 5.14) as a benchmark for embodied cooperation studies. This framework is designed to be applied to both homogeneous and heterogeneous multi-agent systems, with the potential for expansion. Furthermore, previous work did not account for spatiotemporal features in sequences of events, which hinders the understanding of interactive and cooperative behaviors. To overcome this, we introduce sequential behavior learning with spatiotemporal memory in the middle layer of the hierarchical network, using a Recurrent Neural Network (RNN).

4.2 Related Works

4.2.1 Embodied Cooperation

Research on cooperation can be broadly categorized based on whether body interaction is involved. Some collaborative tasks, such as those in particle games, focus solely on coordination and alignment at the position/velocity levels [22, 41]. In contrast, studies on embodied cooperation are grounded in robotic platforms that possess physical bodies, with an emphasis on interaction, collision avoidance, and whole-body control.

Various optimization and computational intelligence methods have been applied to solve embodied cooperation problems. Model Predictive Control (MPC) methods are widely used in formation control for quadruped robots [17, 21, 38, 39] and aerial robots [27, 29]. Additionally, nature-inspired algorithms have demonstrated strong performance in self-organized swarm cooperative tasks [12]. While these approaches provide highly accurate control, they often lack flexibility and generalization. Learning-based methods can address these limitations [5, 25]. Furthermore, large language models (LLMs) have the potential to equip robots with powerful cognitive reasoning capabilities for collaboration, offering a novel approach to solving these challenges [7, 19, 31, 43].

4.2.2 HRL in Locomotion Control

HRL is particularly well-suited for complex, spatiotemporal tasks [26]. Its effectiveness has been widely demonstrated and recognized [32]. By breaking down intricate tasks into multiple steps, HRL reduces training complexity and enables the use of different policy layers [6]. Decisions between these layers can be based on whether the previous action has been completed [40]. In essence, HRL provides structured guidance, reducing the need for a model to fully understand the scenario and the tasks, and preventing reinforcement learning from getting lost in a sparse search space.

Furthermore, HRL is inherently well-suited for tasks that involve a progressive logical relationship [18, 36]. A typical approach combines simplified environmental observations with proprioceptive state information to generate lower-level control commands for actuators [10]. Recent benchmarks, such as HumanoidBench [30], also adopt a hierarchical approach by dividing tasks into planning and reaching control, enabling learning at different levels.

4.3 Problem Statement

We begin by formulating the problem of learning embodied cooperative behavior in a swarm of agents and then attempt to solve it using multi-agent reinforcement learning (MARL). Since the number of agents remains fixed within each episode, we select a synchronous framework and policy gradient descent algorithms for training. To enable real-time execution, we further propose a fully decentralized training framework to support our implementation. It is important to note that the agents in the swarm may be homogeneous or heterogeneous, with agents of the same species sharing the same policy.

The embodied cooperation problem can be formulated as a Markov Decision Process (MDP) mathematically expressed as $\langle \mathcal{S}, \mathcal{A}, \mathcal{T}, \mathcal{R} \rangle$ including N agents in M heterogeneous species. $\mathcal{S}, \mathcal{A}$ are state and action tuples, respectively, where

$\mathcal{S} = (S_1, S_2, \cdots, S_N)$, $\mathcal{A} = (A_1, A_2, \cdots, A_N)$. $\mathcal{T}$ is the state transition function, and $\mathcal{T}(s'|s, a)$ represents the transition probability distribution of the system when transiting into the next states $s' = (s'_1, s'_2, \cdots, s'_N) \in \mathcal{S}$ from states $s = (s_1, s_2, \cdots, s_N) \in \mathcal{S}$ after taking a set of actions $a = (a_1, a_2, \cdots, a_N) \in \mathcal{A}$. Besides, $\mathcal{R} = (R_1, \cdots, R_M)$ denotes the tuple of reward functions in which different types of agents correspond to distinct reward functions. $r_i = R_i(s_i, a_i)$ is the reward given to agent i after taking actions a_i under states s.

For the homogenous multiple agents group, the reward function R is determined and shared for each agent. The aim is to find the best shared policy π^*, to maximize the expected average value of cumulative discount reward:

$$\pi^* = \underset{\pi}{\operatorname{argmax}} J_\pi = \frac{1}{N} \sum_{i=1}^{N} \sum_{t=0}^{\infty} \gamma^t \cdot \mathbb{E}\left[R_i(s_t^i, a_t^i)|\pi_i\right], \tag{4.1}$$

where $s_t^i \in S_i$ and $a_t^i \in A_i$ are the state and action of the i-th agent. There are some subtle differences in the representation of the heterogeneous population. The aim is to find the best policy tuple $\Omega^* = (\pi_1, \cdots, \pi_M)$ to maximize the joint multiplication of costs J:

$$\Omega^* = \underset{\Omega}{\operatorname{argmax}} \prod_{m=1}^{M} J_m, \text{ where } J_m = \frac{1}{K} \sum_{i=1}^{K} \sum_{t=0}^{\infty} \gamma^t \cdot \mathbb{E}\left[R_m(s_t^i, a_t^i)|\pi_m\right], \tag{4.2}$$

where J_m denotes the cost function of type m species, and K is the agent number of this species. The final optimization cost for the heterogeneous population is a form of co-multiplication because the task assignment exists among the entire team. Each subtask must be completed; otherwise, the overall task will fail. Even if some subtasks obtain high rewards, any failure in a subtask will prevent the achievement of the overall goal.

4.4 Distributed Hierarchical Reinforcement Learning

As mentioned earlier, locomotion control for embodied cooperation presents several challenges. These can be summarized into three key difficulties: (i) *How can complex tasks be decomposed and decoupled into step-by-step processes?* (ii) *Can the trained models scale to support large-scale multi-agent collaboration?* (iii) *How can temporal and spatial correlations be established among collective behaviors?* In this section, we will address each of these challenges in detail.

4.4.1 Task Decomposition and Hierarchical Learning

Network Structure Traditionally, hierarchical neural network structures divide tasks into control and decision layers. In contrast, to address collaborative tasks, we propose a deeper hierarchical structure consisting of three layers. The upper layer is responsible for perception-related tasks, extracting exteroceptive observations, including information from other agents, into a feature vector. The middle layer serves as the planning and decision layer, generating spatiotemporally correlated commands for the lower-level controller. Finally, the lower layer focuses on executing these commands, which can be pretrained in various modes. The overall information flow within the proposed hierarchical network is illustrated in Fig. 4.2.

Hierarchical Perception Processing Each layer of the network is responsible for processing different types of raw data, which corresponds to how we acquire and classify perceptual information. Proprioceptive perceptual states, such as local position, velocity, and joint torque, are closely tied to personal locomotion control. Therefore, the proprioceptive state p should only be processed by the lower layer policy. On the other hand, the exteroceptive state e is formed by concatenating environmental perceptions, which provide terrain information in the form of a height measurement array [9], with relative observations of other agents [15], as shown in Fig. 4.3.

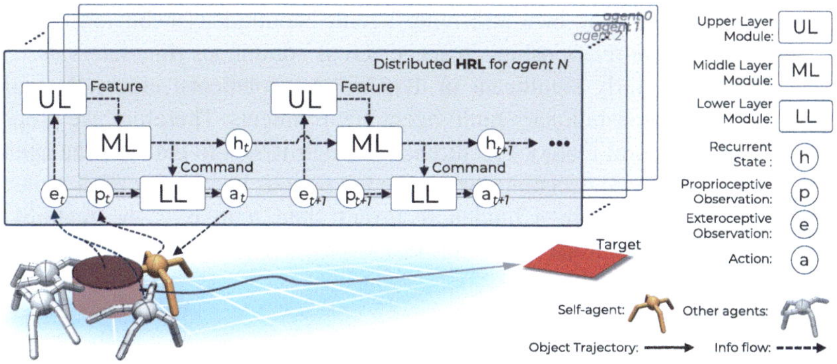

Fig. 4.2 Distributed HRL pipeline. (Reproduced with permission from [11])

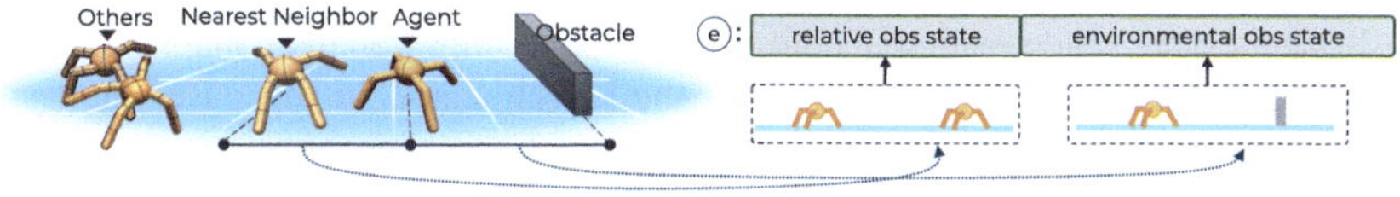

Fig. 4.3 Exteroceptive state configuration in distributed HRL. (Reproduced with permission from [11])

4.4.2 Distributed Scalable HRL

In contrast to previous studies that employed HRL or end-to-end approaches for robotic cooperative manipulation or locomotion, our distributed HRL method takes into account the scalability of the agent population. To achieve this, we must strictly define the format of input and output information for each agent, particularly the observation. In this chapter, we propose using partial observations of agents related to their colleagues, based on their visibility [1]. Each agent can only obtain the relative position of its nearest neighbor. As illustrated in Fig. 4.2, the dimension of the exteroceptive state e is limited. This limitation alters the difficulty of collaborative learning. Since partially observed agents can only perceive nearby peers and environmental information, they must estimate the states of other invisible agents from the dynamics of the environment and predict future changes in it. However, such observation settings theoretically enable our framework to scale to significantly larger agent systems, a capability unattainable by CTDE frameworks.

4.4.3 Spatiotemporal Memory Recurrence

Capturing and leveraging spatiotemporal features is crucial for long-term decision-making tasks [44], especially for the complex embodied cooperation tasks we designed. We anticipate that fully decentralized agents should possess the ability to comprehend both their own and other agents' continuous behaviors, thereby forming collaborative relationships in space across continuous time intervals. This capability is particularly significant in dynamic environments, especially within fully decentralized nonstationary multi-agent environments. Therefore, we propose utilizing spatiotemporal memory recurrence through RNNs to enhance the agent's ability to make informed decisions over extended periods in complex environments. Consequently, we maintain a hidden recurrent state h to pass the features of scenarios continuously, as shown in Fig. 4.2.

4.4.4 Training Curriculum

Acquiring collaborative behaviors requires the agents' random search to span a sufficiently large space, which in complex tasks can often lead to an inability to learn usable knowledge due to the vastness of the search space. In this case, we design some appropriate training curricula that reasonably use sparse rewards and dense rewards in the early stages to intentionally guide the agent to approach success. The curriculum designs are described in detail in the experiments part.

4.5 Experiments

The experiments are conducted on a workstation with Intel Xeon W-2150b and RTX2080Ti. Independent Proximal Policy Optimization (IPPO) in a multi-agent setting is employed as the algorithm. Adam optimizer is used with a learning rate of 0.0005. PPO clipping is 0.2, and the discount factor is 0.995. We could complete the training within several hours thanks to IsaacSim GPU parallel sampling via multiple environments [23].

Firstly, we showcase the details of the reward mechanism of the lower layer that controls the fundamental movement of the agent. Herein, we use two kinds of control paradigms: target control and velocity control (Tables 4.1 and 4.2).

4.5.1 Environments Construction

Three scenarios are established that represent different task types: Cooperative Transport, Corridor Crossing, and Ravine Bridging.

Table 4.1 Reward details for target paradigm on lower layer

Reward type	Formula
Distance to target reward	$5 \times \left(\frac{1}{1+\sqrt{(\text{agent\_pos}[:,0]-\text{target\_pos}[:,0])^2+(\text{agent\_pos}[:,1]-\text{target\_pos}[:,1])^2}} \right)$
Alive reward	0.5
Up reward	0.1
Action cost	$0.005 \times$ action cost
Energy cost	$0.05 \times$ energy cost
Dof at limit cost	$0.1 \times$ dof at limit cost

Table 4.2 Reward details for velocity paradigm on lower layer

Reward type	Formula
Vel reward	$5 \times \text{clamp}(\text{vel}[:, : 2] \cdot \text{cmd\_direction}, \text{max} = 1)$
Alive reward	0.5
Up reward	0.1
Action cost	$0.005 \times$ action cost
Energy cost	$0.05 \times$ energy cost
Dof at limit cost	$0.1 \times$ dof at limit cost

4.5.1.1 Cooperative Transport

Transporting an object collaboratively is a classical task in the multi-agent system field. Herein, we build an embodied cooperative transport task via locomotion control. A group of `Ant` attempts to move a cylinder object to the red target area, shown in Fig. 5.14a. The target position is initialized randomly. The cylinder is set in the middle of the map, and agents are initialized in the four quadrants of the cylinder, respectively. We consider fully decentralized and distributed cooperation in a homogeneous and self-organized flock. Thus, the scale of the flock can theoretically expand without limit.

Observation The agents receive two sets of observations: (i) a set of "proprioceptive" states which would be given to the lower layer controller. This part of the observation is exactly the same in all agents in different tasks because all tasks use the same pretrained lower layer policy, including their self-observed information, such as its local position, local velocity, angular velocity, orientation, body upright angle, joint force, and its limitations, etc. and (ii) a set of "exteroceptive" features containing task-relevant information which would be given to upper layer controller. In this task, the observation includes the locations of the nearest agent, the current position of the cylinder, and the destination.

Rewards We set three types of rewards: (i) Cylinder distance reward: A term which is the reciprocal L2 distance of the agent to the target. This is the main reward of this task. It will increase when the cylinder becomes closer to the target. The increasing trend is the same as the inverse proportional function. (ii) Agent cylinder distance reward: This term is to encourage agents firstly to come closer to the cylinder so that they can learn how to push the object. (iii) Agent distance reward: A term used to guide the agent to walk toward the target, thereby obtaining a large cylinder distance reward (Table 4.3).

Table 4.3 Reward details for cooperative transport on upper layer

Reward type	Formula
Cylinder-target distance reward	$10 \times \left(\frac{1}{1+\sqrt{(\text{cylinder\_pos}[:,0]-\text{target\_pos}[:,0])^2+(\text{cylinder\_pos}[:,1]-\text{target\_pos}[:,1])^2}} \right)$
Agent-cylinder distance reward	$5 \times \left(\frac{1}{1+\sqrt{(\text{agent\_pos}[:,0]-\text{cylinder\_pos}[:,0])^2+(\text{agent\_pos}[:,1]-\text{cylinder\_pos}[:,1])^2}} \right)$
Agent-target distance reward	$5 \times \left(\frac{1}{1+\sqrt{(\text{agent\_pos}[:,0]-\text{target\_pos}[:,0])^2+(\text{agent\_pos}[:,1]-\text{target\_pos}[:,1])^2}} \right)$
Reach destination reward	50 (all agents)

4.5.1.2 Corridor Crossing

A group of agents navigate through a narrow corridor sequentially, aiming to complete the passage as quickly as possible. Only one agent can pass through the slit at any given time. However, all agents are at the same distance from the entrance of the slit, meaning that if no agent slows down or stops to allow others to pass, all agents would become stuck at the entrance. We expect that agents will acquire a strategy such that they can all pass through the corridor as quickly as possible. In this scenario, agents must adjust their position and speed based on surrounding observations and pass through the narrow passage without collision. Cooperation information is required, as shown in Fig. 5.14b. The number of agents can also be expanded in this task.

Observation The observation for the lower layer controllers is the same as in the last task. In this task, the upper layer controllers' observation includes the location of the agent itself and its nearest neighbor, as well as the position, translation, and orientation of the walls.

Rewards Two rewards are set: (i) Destination reward: If an agent has passed the finish line, all agents will receive a positive reward. It stimulates the strategy allowing as many agents as possible to pass through the corridor to the finish line. (ii) Distance reward to enter: A term to encourage agents first to come close to the corridor (Table 4.4).

4.5.1.3 Ravine Bridging

The agents are expected to reach a designated line, where they must first cross over a ravine. A ravine blocks the agents' path, and they cannot jump over or climb out of it. The agents need the help of their colleagues to get across the ravine. To address this, we place a stone of the appropriate size and depth in the ravine to serve as a "bridge" to help the agents cross. We preset two agents on either side of the stone to push it and help their colleagues cross the ravine and reach the destination. This is a complex task that requires teamwork among heterogeneous agents.

As illustrated in Fig. 5.14c, two agents push a movable bridge, enabling another group of agents to cross the chasm smoothly via the bridge. There are two categories of agents in this scenario, which is different from the other two environments.

Table 4.4 Reward details for cross corridor on upper layer

Reward type	Formula
Reach destination reward	50 (all agents)
Agent-target distance reward	$5 \times \left(\frac{1}{1+\sqrt{(\text{agent\_pos}[:,0]-\text{target\_pos}[:,0])^2+(\text{agent\_pos}[:,1]-\text{target\_pos}[:,1])^2}} \right)$

Table 4.5 Reward details for ravine bridging on upper layer

Reward	Formula
Ravine pretrain: Distance of stone and nearest agent reward	$5 \times \left(\frac{1}{1+\lvert stone\_y-agent\_y\rvert}\right)$
Plane pretrain: Distance to destination line reward	$5 \times \left(\frac{1}{1+\lvert destination\_x-agent\_x\rvert}\right)$
Plane pretrain: Falling down punishment	– 20
Integrated training: Pass destination line reward	50 (all agents)

Agents in `Ravine Bridging` are heterogeneous, which means we should maintain more than one policy for different missions.

Observation The observation of the lower layer policy is similar to the previous tasks. Agents' upper layer controllers' observation includes the locations of itself and the nearest agents, the position of the stone, and the position and depth of the ravine.

Rewards Two rewards are set: (i) Destination reward: If any agent has passed the finish line, all agents will receive a positive reward. It stimulates the strategy allowing as many agents as possible to go through the ravine and reach the finish line. (ii) Bridging reward: A term to encourage agents in the ravine to carry the stone to the agent above which is the closest to the ravine (Table 4.5).

4.5.2 *Results of Distributed HRL*

4.5.2.1 Training Basic Locomotion Control

In the training of lower-level motion control, two control paradigms are available: velocity control and position control. We first train the two control paradigms in the single-agent environment to obtain accurate control of the fundamental actions. The main reward in each training curriculum is given as the reciprocal L_2 distance between the current position and the target position, or the dot product of the agent's velocity and the target velocity. Figure 4.4a shows the training results of the two paradigms. After training, we find that velocity control offers greater flexibility than position control. However, position control presents higher accuracy. Moreover, velocity control makes it difficult to maintain a stationary position at a specific point.

4.5.2.2 Training the Interactive Collaboration

After the lower layer is elaborately trained, we adopt the position control paradigm and freeze this network. Then, the upper and middle layers are trained on different scenes via distinct reward mechanisms. The training results are shown in Fig. 4.4b,c,d.

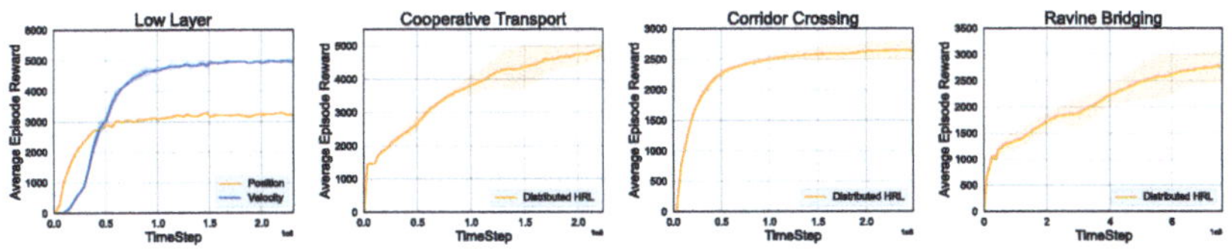

(a) (b) (c) (d)

Fig. 4.4 Distributed HRL results for all scenarios. (Reproduced with permission from [11]). (**a**) Basic locomotion. (**b**) Cooperative transport. (**c**) Corrider crossing. (**d**) Ravine bridging

Cooperative Transport This task reveals clear collaborative behaviors among agents. They demonstrate a strong understanding of the situation, actively choosing tasks based on their circumstances. Some agents first converge toward the object and slightly adjust their position for collaboration; meanwhile, other agents move directly toward the target and prepare to decelerate the object, ensuring it falls within the target area accurately and stably. The success rate of this task reaches approximately 75%.

Corridor Crossing In this task, agents naturally develop behaviors such as "sacrifice" or "concession." If a model pretrained in a single-agent scenario is directly applied to a multi-agent setting, all agents would get stuck in the corridor and fail to progress. However, after training, agents learn to observe the relative positions of others and adjust their speed accordingly, carefully timing their movements and pauses. Typically, agents in the middle of the corridor move quickly, while those at the entrance slow down and accelerate after observing the agents ahead. The success rate of this task is approximately 88%.

Ravine Bridging The group in this task is a heterogeneous team. Although all agents are `Ant`, there are two sets of policies: one for agents on the plane and one for agents in the ravine. Agents on the plane will arrive at the edge of the ravine and stop, waiting for the agents in the ravine to transport the bridge to the front of them. The success rate of this task reaches approximately 50%. The main reason is that `Ants` in the ravine might get stuck or fall.

4.5.3 Ablation

4.5.3.1 The Role of Hierarchical Learning

In this experiment, we replace the hierarchical networks with a single-layer MLP. Without the hierarchical structure, there are no target or velocity commands, making it impossible to pretrain an identical lower-level controller. Therefore, when training without hierarchy, we apply the same training curriculum tailored to each task, as detailed in the environment construction section. Additionally, we provide both

Table 4.6 Success rate comparison: We present success rates with standard error on the three tasks, comparing our method to ablations. Each percentage results from 500 trials—100 attempts per five independently trained models

Success rate	HRL	No hierarchy	No spatiotemporal memory
Cooperative transport	**76.8 ± 5.3%**	0.0 ± 0.0%	9.6 ± 3.2%
Corridor crossing	**88.4 ± 4.8%**	0.0 ± 0.0%	12.0 ± 4.1%
Ravine bridging	**50.4 ± 4.4%**	0.0 ± 0.0%	5.4 ± 4.2%

proprioceptive and exteroceptive states as observations, with the same task and locomotion rewards serving as the total rewards.

We find that without hierarchy, agents are only able to learn simple strategies and achieve basic rewards, which are easily attainable. These results highlight the importance of hierarchy for success in more complex tasks requiring coordination. Without a hierarchical structure, agents fail to complete the mission, as shown in Table 4.6. The reason for this failure is clear: Similar to other HRL objectives, a single network facing high-dimensional observations and complex, multi-item reward functions struggles to distinguish features across channels and learn behavior driven by a single, primary reward. Notably, the absence of a pretrained lower-level locomotion controller, which is essential for success, exacerbates this issue.

4.5.3.2 The Effectiveness of Spatiotemporal Memory

To assess the effectiveness of spatiotemporal features, we replicate the method from previous work [25] as a baseline, which does not account for event correlation. We provide the same low-level locomotion controller for consistency. The approach proposed in [25], which lacks RNNs, converges to a slightly higher reward compared to the no-hierarchy case but still falls far short of mission completion. As shown in Table 4.6, the success rate of this method remains relatively low, rarely exceeding 15%. We observe that agents are unable to learn a systematic set of behaviors for specific tasks, with occasional successes occurring randomly. In fact, when agents only react to current observations, they fail to develop strategic behaviors. Without spatiotemporal memory, agents cannot anticipate future steps. This limitation is particularly detrimental in collaborative tasks, where we see that the cooperative and sacrificial behaviors emerging in the corridor task are essential for long-term strategy and success.

4.5.4 Scalability Experiments

We highlight the advantages of our proposed fully decentralized approach, particularly its scalability. One key benefit is its ability to handle large numbers of agents during training, as we do not require global observations of all agents, unlike the

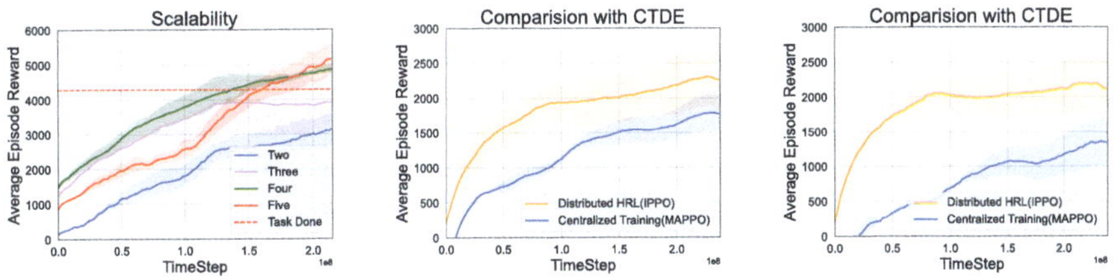

(a) (b) (c)

Fig. 4.5 (**a**) shows the training results of different `Ant` populations used in Cooperative Transport. (**b**), (**c**) shows the result training with large agent numbers. (Reproduced with permission from [11]). (**a**) Training results for different `Ant` populations. (**b**) Training with six agent number. (**c**) Training with seven agent number

Table 4.7 Success rate of each task with different populations

Success rate	Two ants	Three ants	Four ants	Five ants
Cooperative transport	$12.8 \pm 5.6\%$	$39.8 \pm 4.5\%$	$\mathbf{76.8 \pm 5.3\%}$	$82.2 \pm 4.5\%$
Corridor crossing	$97.1 \pm 1.8\%$	$93.6 \pm 2.2\%$	$\mathbf{88.4 \pm 4.8\%}$	$73.5 \pm 2.5\%$
Ravine bridging	$\mathbf{50.4 \pm 4.4\%}$	$36.2 \pm 7.8\%$	$22.3 \pm 5.1\%$	$10.9 \pm 3.3\%$

CTDE framework, which significantly increases training complexity. Additionally, our framework offers the advantage of seamlessly transferring the trained model to tasks involving more agents, ensuring greater flexibility and efficiency in multi-agent environments.

First, we demonstrate the capability of our framework to effectively train agents and achieve strong performance as the number of agents increases, even when they only have access to partial observations. We validate this in the cooperative transport task, as illustrated in Fig. 4.5a. The final reward increases with the number of agents, which aligns with our expectations for this task. As more agents participate, the transportation results improve, confirming the scalability and effectiveness of our approach.

Second, we conducted zero-shot transfer tests on our trained model (initially trained with four agents) across various populations of `Ant` agents in each task (with the agent count indicating the number of agents on the plane in the Ravine Bridging task). The success rates for each task with different agent populations are shown in Table 4.7. The results highlight the transferability and scalability of our framework.

Third, we conducted an experiment to demonstrate the advantages of our fully decentralized method over CTDE methods as the number of agents increases. The Corridor Crossing task is particularly suitable for this comparison, as the difficulty of the Cooperative Transport task decreases with more agents, while Ravine Bridging becomes too complex, with a low completion rate when the agent count is large. We tested with six and seven agents in the Corridor Crossing task. As shown in Fig. 4.5b,c, the reward for centralized training increases more slowly and ultimately fails to converge.

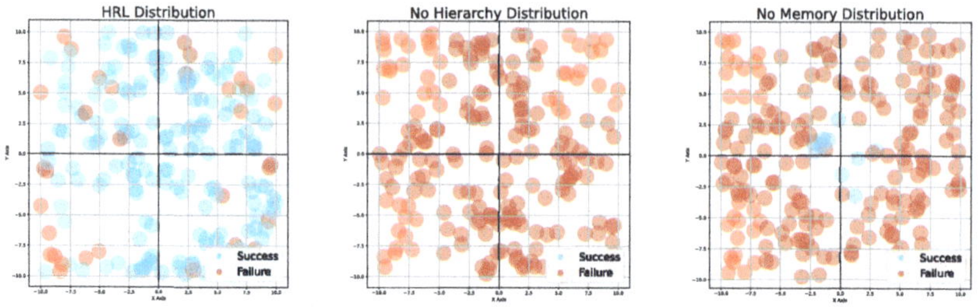

(a) (b) (c)

Fig. 4.6 (**a**) (**b**) (**c**) show the success and failure distribution of task Cooperative Transport. (Reproduced with permission from [11]). (**a**) Distributed HRL. (**b**) No hierarchy. (**c**) No memory

4.5.5 Typical Result

4.5.5.1 Cooperative Transport

We analyze the typical results of the Cooperative Transport task to assess the robustness of our method. To do this, we conduct over one hundred trials with randomized reset positions and calculate the success rate, which we then visualize in a scatter plot, as shown in Fig. 4.6b. Each point represents the initial position of an agent, with the object located at the (0,0) point. Red points indicate positions where agents failed to complete the task, while blue points denote successful completions. We observe that success rates tend to be higher when the initial position is closer to the object, with only one failure case occurring within a 5-meter range of the object. Conversely, as the distance from the object increases, the success rate decreases, likely due to a higher likelihood of accidents in longer-distance tasks. Notably, most failures arise from agents tipping over or pushing the object with uneven force, causing it to veer off to the side. Figure 4.6c, d presents the evaluation results for the no-hierarchy and no-spatiotemporal memory conditions. In the no-spatiotemporal memory scenario, only a few successes are observed, while all cases fail when training without hierarchy.

4.5.5.2 Corridor Crossing

We record the success rates of each `Ant` at its initial position separately, as illustrated in Fig. 4.7. We have a higher success rate when the initial position is closer to the center, and a higher failure rate when the initial position is closer to the edge. This is also explainable, as the `Ant` at the center is closer to the slit and does not need to adjust velocity (Fig. 4.8).

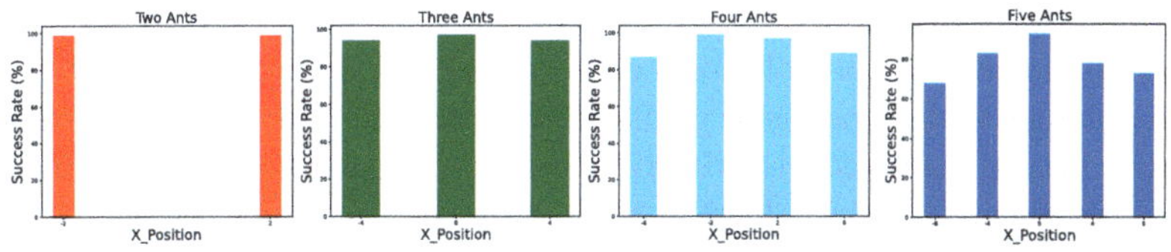

Fig. 4.7 Success rate for Corridor Crossing. (Reproduced with permission from [11])

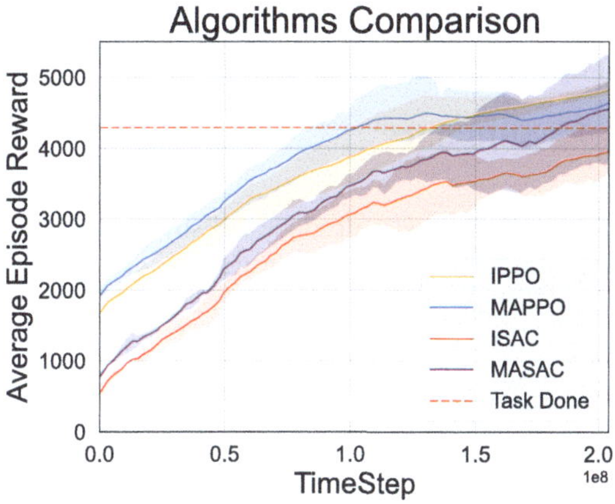

Fig. 4.8 Results of D-HRL in different algorithms. (Reproduced with permission from [11])

4.5.6 Comparison to Other MARL Algorithms

We conduct the comparison between the proposed distributed HRL method and other state-of-the-art MARL approaches based on the article [2]. Specifically, we compare our fully decentralized method with MAPPO and MASAC in task Cooperative Transport (four agents) since the Isaac Sim algorithm is nested within the RLGame, and the CTDE algorithms supported by RLGame are only MAPPO and MASAC. Figure 4.7 shows the results. We find that MAPPO converges a little faster than the distributed method because it is centrally trained with global observations. However, the CTDE algorithm has worse performance and takes much more training time when facing a larger number of agents, which has been demonstrated in Fig. 4.5 of the scalability experiments part.

4.6 Limitations

Despite the promising results, it is important to acknowledge the limitations of our approach. First, while the hierarchical network method provides our framework with some capability to address long-horizon problems, its ability to tackle more complex issues is still limited, as all layers are fully trained using reinforcement learning. We primarily relied on continuous reward shaping and an increase in the total training steps to achieve improved training outcomes. However, we believe that in order to solve more complex, long-horizon cooperative tasks involving more intelligent agents, additional planning-based guidance at the upper level may be necessary.

Moreover, this approach is primarily focused on algorithmic and simulation-based benchmarks, and we have not yet explored robots with more complex morphologies. Although we have demonstrated promising results on certain tasks and observed the emergence of cooperative behavior, there remains a significant engineering workload, particularly in terms of control and sim-to-real transfer.

4.7 Summary

In this chapter, we introduce a distributed HRL architecture for embodied cooperation, and our results demonstrate its effectiveness in collaborative tasks involving embodied and whole-body control. Additionally, the findings highlight the scalability potential of our distributed HRL framework in zero-shot and one-shot transfer scenarios, which is essential for training large-scale multi-agent systems. Furthermore, the adaptability of our architecture is evidenced by its performance across diverse task environments, underscoring its potential for practical applications in areas such as robot swarms, autonomous drone fleets, and collaborative robotics in dynamic environments.

References

1. Amorim T, Nascimento T, Petracek P, De Masi G, Ferrante E, Saska M (2021) Self-Organized UAV flocking based on proximal control. In :2021 International conference on unmanned aircraft systems, ICUAS 2021 pp 1374–1382. https://doi.org/10.1109/ICUAS51884.2021.9476847
2. Bettini M, Prorok A, Moens V (2024) BenchMARL: Benchmarking multi-agent reinforcement learning. arXiv preprint arXiv:2312.01472. https://arxiv.org/abs/2312.01472
3. Cao J, Dong L, Sun C (2024) Hierarchical reinforcement learning for kinematic control tasks with parameterized action spaces. Neural Comput Appl 36(1):323–336. https://doi.org/10.1007/s00521-023-08991-2
4. Chang L, Shan L, Zhang W, Dai Y (2023) Hierarchical multi-robot navigation and formation in unknown environments via deep reinforcement learning and distributed optimization. Robot Comput Integr Manuf 83:102570. https://doi.org/10.1016/j.rcim.2023.102570
5. De Witte S, Van Hauwermeiren T, Lefebvre T, Crevecoeur G (2022) Learning to cooperate: a hierarchical cooperative dual robot arm approach for underactuated pick-and-placing. IEEE/ASME Trans Mechatron 27(4):1964–1972. https://doi.org/10.1109/TMECH.2022.3175484
6. Eppe M, Gumbsch C, Kerzel M, Nguyen PD, Butz MV, Wermter S (2022) Intelligent problem-solving as integrated hierarchical reinforcement learning. Nat Mach Intell 4(1):11–20. https://doi.org/10.1038/s42256-021-00433-9
7. Guo X, Huang K, Liu J, Fan W, Vélez N, Wu Q, Wang H, Griffiths TL, Wang M (2024) Embodied LLM agents learn to cooperate in organized teams. http://arxiv.org/abs/2403.12482

8. Hao C, Weaver C, Tang C, Kawamoto K, Tomizuka M, Zhan W (2024) Skill-Critic: refining learned skills for hierarchical reinforcement learning. IEEE Rob Autom Lett 9(4):3625–3632. https://doi.org/10.1109/LRA.2024.3368231
9. Heess N, TB D, Sriram S, Lemmon J, Merel J, Wayne G, Tassa Y, Erez T, Wang Z, Eslami SMA, Riedmiller M, Silver D (2017) Emergence of locomotion behaviours in rich environments. http://arxiv.org/abs/1707.02286
10. Heess N, Wayne G, Tassa Y, Lillicrap T, Riedmiller M, Silver D (2016) Learning and transfer of modulated locomotor controllers. http://arxiv.org/abs/1610.05182
11. Hong C, Huang K, Liu H: Learning a distributed hierarchical locomotion controller for embodied cooperation. In: 8th Annual Conference on Robot Learning (2024). https://openreview.net/forum?id=NCnplCf4wo
12. Huang K, Chen J, Oyekan J (2021) Decentralised aerial swarm for adaptive and energy efficient transport of unknown loads. Swarm Evol Comput 67. https://doi.org/10.1016/j.swevo.2021.100957
13. Huang K, Chen J, Oyekan J, Zhang X (2022) Bio-inspired multi-agent model and optimization strategy for collaborative aerial transport. In: Lecture Notes in Electrical Engineering, vol 801 LNEE, pp 591–598. https://doi.org/10.1007/978-981-16-6372-7_64
14. Huang K, Guo D, Zhang X, Ji X, Liu H (2024) CompetEvo: towards morphological evolution from competition. http://arxiv.org/abs/2405.18300
15. Huang K, Guo D, Zhang X, Ji X, Liu H (2024) Stimulate the potential of robots via competition. http://arxiv.org/abs/2403.10487
16. Huang X, Li Z, Xiang Y, Ni Y, Chi Y, Li Y, Yang L, Peng XB, Sreenath K (2023) Creating a dynamic quadrupedal robotic goalkeeper with reinforcement learning. In: IEEE International Conference on Intelligent Robots and Systems, pp 2715–2722. https://doi.org/10.1109/IROS55552.2023.10341936
17. Kim J, Fawcett RT, Kamidi VR, Ames AD, Hamed KA (2023) Layered control for cooperative locomotion of two quadrupedal robots: centralized and distributed approaches. IEEE Trans Robot 39(6):4728–4748. https://doi.org/10.1109/TRO.2023.3319896
18. Li T, Lambert N, Calandra R, Meier F, Rai A (2020) Learning generalizable locomotion skills with hierarchical reinforcement learning. In: Proceedings—IEEE international conference on robotics and automation, pp 413–419. https://doi.org/10.1109/ICRA40945.2020.9196642
19. Liu X, Guo D, Zhang X, Liu H (2024) Heterogeneous embodied multi-agent collaboration. IEEE Rob Autom Lett 9(6):5377–5384. https://doi.org/10.1109/LRA.2024.3390588
20. Liu X, Wang G, Liu Z, Liu Y, Liu Z, Huang P (2023) Hierarchical reinforcement learning integrating with human knowledge for practical robot skill learning in complex multi-stage manipulation. In: IEEE transactions on automation science and engineering. https://doi.org/10.1109/TASE.2023.3288037
21. Ma WL, Csomay-Shanklin N, Kolathaya S, Hamed KA, Ames AD (2021) Coupled control lyapunov functions for interconnected systems, with application to quadrupedal locomotion. IEEE Rob Autom Lett 6(2):3761–3768. https://doi.org/10.1109/LRA.2021.3065174
22. Ma X, Yang Y, Li C, Lu Y, Zhao Q, Yang J (2021) Modeling the interaction between agents in cooperative multi-agent reinforcement learning. In: Proceedings of the international joint conference on autonomous agents and multiagent systems, AAMAS, vol. 2, pp 853–861
23. Makoviychuk V, Wawrzyniak L, Guo Y, Lu M, Storey K, Macklin M, Hoeller D, Rudin N, Allshire A, Handa A, State G (2021) Isaac gym: high performance GPU-based physics simulation for robot learning. http://arxiv.org/abs/2108.10470
24. Mohanty S, Nygren E, Laurent F, Schneider M, Scheller C, Bhattacharya N, Watson J, Egli A, Eichenberger C, Baumberger C, Vienken G, Sturm I, Sartoretti G, Spigler G (2020) Flatland-RL: multi-agent reinforcement learning on trains. http://arxiv.org/abs/2012.05893
25. Nachum O, Ahn M, Ponte H, Gu S, Kumar V: Multi-agent manipulation via locomotion using hierarchical Sim2Real. In: Proceedings of machine learning research, vol 100, pp 110–121 (2019)
26. Nachum O, Lee H, Gu S, Levine S (2018) Data-efficient hierarchical reinforcement learning. Adv Neural Inf Proces Syst 2018-Decem:3303–3313

27. Papaioannou S, Kolios P, Panayiotou CG, Polycarpou MM (2020) Cooperative simultaneous tracking and jamming for disabling a rogue drone. In: IEEE international conference on intelligent robots and systems, pp 7919–7926. https://doi.org/10.1109/IROS45743.2020.9340835
28. Parvini M, Javan MR, Mokari N, Abbasi B, Jorswieck EA (2023) AoI-aware resource allocation for platoon-based C-V2X networks via multi-agent multi-task reinforcement learning. IEEE Trans Veh Technol 72(8):9880–9896. https://doi.org/10.1109/TVT.2023.3259688
29. Quan L, Yin L, Zhang T, Wang M, Wang R, Zhong S, Zhou X, Cao Y, Xu C, Gao F (2023) Robust and efficient trajectory planning for formation flight in dense environments. IEEE Trans Robot 39(6):4785–4804. https://doi.org/10.1109/TRO.2023.3301295
30. Sferrazza C, Huang DM, Lin X, Lee Y, Abbeel P: HumanoidBench: simulated humanoid benchmark for whole-body locomotion and manipulation (2024). http://arxiv.org/abs/2403.10506
31. Sun C, Huang S, Pompili D (2024) LLM-based multi-agent reinforcement learning: current and future directions. http://arxiv.org/abs/2405.11106
32. Sun C, Liu W, Dong L (2021) Reinforcement learning with task decomposition for cooperative multiagent systems. IEEE Trans Neural Networks Learn Syst 32(5):2054–2065. https://doi.org/10.1109/TNNLS.2020.2996209
33. Triantafyllidis E, Acero F, Liu Z, Li Z (2023) Hybrid hierarchical learning for solving complex sequential tasks using the robotic manipulation network ROMAN. Nat Mach Intell 5(9):991–1005. https://doi.org/10.1038/s42256-023-00709-2
34. Wang R, Zhao D, Gupte A, Min BC (2024) Initial task allocation in multi-human multi-robot teams: an attention-enhanced hierarchical reinforcement learning approach. IEEE Rob Autom Lett 9(4):3451–3458. https://doi.org/10.1109/LRA.2024.3366414
35. Wang S, Wu Z, Hu X, Lin Y, Lv K (2023) Skill-based hierarchical reinforcement learning for target visual navigation. IEEE Trans Multimedia 25:8920–8932. https://doi.org/10.1109/TMM.2023.3243618
36. Wöhlke J, Schmitt F, van Hoof H (2021) Hierarchies of planning and reinforcement learning for robot navigation. In: Proceedings—IEEE International Conference on Robotics and Automation, vol 2021-May, pp 10682–10688. https://doi.org/10.1109/ICRA48506.2021.9561151
37. Xiong Z, Chen B, Huang S, Tu WW, He Z, Gao Y (2024) MQE: Unleashing the power of interaction with multi-agent quadruped environment. arXiv preprint arXiv:2403.16015. https://arxiv.org/abs/2403.16015
38. Xu S, Zhang W, Zhu L, Ho CP (2023) Distributed model predictive formation control with gait synchronization for multiple quadruped robots. In: Proceedings—IEEE international conference on robotics and automation, vol 2023-May, pp 9995–10002. https://doi.org/10.1109/ICRA48891.2023.10161260
39. Yang C, Sue GN, Li Z, Yang L, Shen H, Chi Y, Rai A, Zeng J, Sreenath K (2022) Collaborative navigation and manipulation of a cable-towed load by multiple quadrupedal robots. IEEE Rob Autom Lett 7(4):10041–10048. https://doi.org/10.1109/LRA.2022.3191170
40. Yang X, Ji Z, Wu J, Lai YK, Wei C, Liu G, Setchi R (2021) Hierarchical reinforcement learning with universal policies for multistep robotic manipulation. IEEE Trans Neural Networks Learn Syst 33(9):4727–4741. https://doi.org/10.1109/TNNLS.2021.3059912
41. Yang Y, Luo R, Li M, Zhou M, Zhang W, Wang J (2018) Mean field multi-agent reinforcement learning. In: 35th International Conference on Machine Learning, ICML 2018, vol 12, pp 8869–8886
42. Yu C, Velu A, Vinitsky E, Gao J, Wang Y, Bayen A, Wu Y (2022) The surprising effectiveness of PPO in cooperative multi-agent games. In: Advances in neural information processing systems, vol 35
43. Zhang H, Du W, Shan J, Zhou Q, Du Y, Tenenbaum JB, Shu T, Gan C (2023) Building cooperative embodied agents modularly with large language models. http://arxiv.org/abs/2307.02485
44. Zhao J, Huang F, Lv J, Duan Y, Qin Z, Li G, Tian G (2020) Do RNN and LSTM have long memory? In: 37th international conference on machine learning, ICML 2020, vol PartF168147-15, pp 11302–11312

Chapter 5
Competitive Learning in Embodied Multi-agent System

Abstract Learning from competitive interactions is essential for improving the capabilities of both single- and multi-agent systems. For embodied agents, where physical interactions play a key role, competition becomes more complex, especially when dealing with adversarial behaviors between limbs. To explore how agents can learn from competition, this chapter proposes a competitive learning framework that enables individuals to acquire knowledge through interactive competition. Additionally, an adversarial learning framework is introduced to facilitate the extraction of knowledge from adversarial behaviors. Furthermore, a joint-optimization approach is presented to evolve agent designs and tactics in cooperative interactive confrontations, allowing agents to adapt effectively to their opponents.

5.1 Introduction

Competition in multi-agent system is both intriguing and widely studied. Previous research in this area often focuses on noncooperative algorithms, typically involving simplified particle dynamics [28, 46]. While these works have made significant contributions, the context is somewhat different in embodied multi-agent system, where the emphasis is on the physical presence and interactions of an agent's body.

In this chapter, we focus on the competitive learning of actions, movements, and locomotor behaviors among agents with physical bodies. We begin by posing several key questions: (i) How can agents learn interactively from raw competitive data? (ii) How can adversarial learning be utilized to enhance robotic capabilities? (iii) How can agents continuously refine their skills and even train their bodies through interaction to better adapt to tasks? We will explore these questions and attempt to provide insights into each of them.

H. Liu et al., *Embodied Multi-Agent Systems*, Machine Learning: Foundations, Methodologies, and Applications, https://doi.org/10.1007/978-981-96-5871-8_5

5.2 Learning from Competitive Information

It is common for us to feel pressure in an interactive competition environment, which arises from the desire to obtain success compared with other individuals or opponents. For example, multiple race athletes often have the ability to achieve better results in competition that exceed their performance in individual training. This might be because athletes can be inspired by pressure from interactive and competitive information.

Learning from competition has the similar nature with contrastive learning, which has been widely used in word and sentence embedding in NLP [11], image classification in CV [8], and implicit collaborative filtering in information retrieval (IR). It is able to extract meaningful representations through positive and negative data pairs. Furthermore, generative adversarial network (GAN) also has drawn significant attention in recent years [9] for selecting negatives, while the confrontation between the generator and discriminator may not converge to the ideal NE, and there is still potential for further exploration and improvement in the adversarial negative sampling method [43]. However, no supervised labels can be obtained from the comparison in reinforcement learning. Thus, how to utilize the competitive information in reinforcement learning?

To further explore the effectiveness of learning from competition, we propose a competitive learning framework that is able to help the individual agent to acquire knowledge from the competitive interaction, fully stimulating its dynamics potential. Specifically, the interactive information among competitors is introduced as the additional auxiliary signal to enhance the learning process. The framework is shown in Fig. 5.2. Our methods are validated on common agents like `Ant`, `HalfCheetah`, `Walker2d`, and `Hopper`, et al., built in most reinforcement learning environments, shown in Fig. 5.1.

5.2.1 *Problem Formulation*

We first formulate the single-agent continuous action task as a Markov Decision Process (MDP), which can be described as a tuple $\langle S, A, P, R \rangle$. S is the state set of a system or environment. A is the action set that the agent can take. P is the state transition function, and $P\left(s'|s, a\right)$ represents the transition probability distribution of the system when transiting into the next state $s' \in S$ from state $s \in S$ after taking an action $a \in A$. R is the reward function. $r = R\left(s, a\right)$ is the reward given to agent after taking action $a \in A$ under $s \in S$. The aim is to find a policy π^*, to maximize the expected value of cumulative discount rewards:

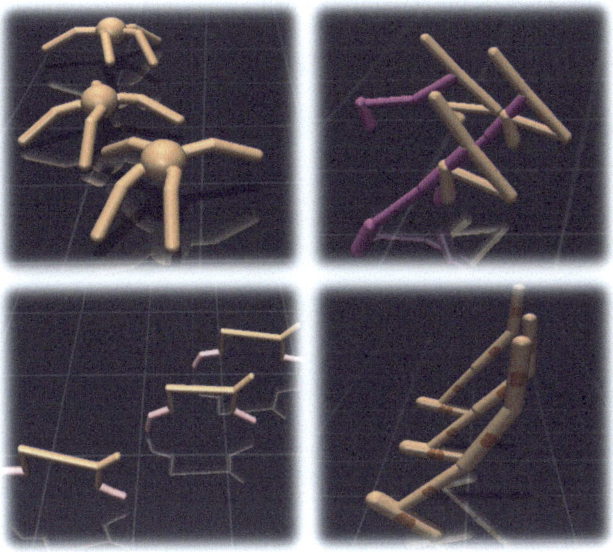

Fig. 5.1 Environments constructed for competitive learning. (Reproduced with permission from [22])

$$\pi^* = \underset{\pi}{\operatorname{argmax}} J_\pi$$
$$J_\pi = \sum_{t=0}^{\infty} \gamma^t \cdot \mathbb{E}\left[R(s_t, a_t) | s_0, \pi\right], \tag{5.1}$$

where γ represents the extent to which the agent discounts future awards.

In this chapter, we consider the multi-agent scenario. We describe the corresponding MDP as a tuple $\langle \mathcal{S}, \mathcal{A}, P, R \rangle$ that includes N number of agents. $\mathcal{S}, \mathcal{A}$ are state and action tuples, respectively, where $\mathcal{S} = (S_1, S_2, \cdots, S_N)$, $\mathcal{A} = (A_1, A_2, \cdots, A_N)$. P and R remain the same as settings in MDP because agents are totally homogeneous, and there is no physical interaction between competitors. In this regard, the objective function in (5.1) can be extended as

$$\hat{J}_\pi = \frac{1}{N} \sum_{i=1}^{N} \sum_{t=0}^{\infty} \gamma^t \cdot \mathbb{E}\left[R(s_t^i, a_t^i) | s_0^i, \pi\right], \tag{5.2}$$

where $s_t^i \in S_i$ and $a_t^i \in A_i$ are the state and action of the i-th agent. However, the above setting only facilitates the parallel exploration but does not provide extra benefits due to the lack of interaction between agents.

In practical training, other agents for the i-th agent can always provide information which can be augmented to the state. We use o_t^i to denote the competitive information which can be observed from other agents and may form a new state as

$$\bar{s}_t^i = [s_t^i, o_t^i]$$

and the objective function becomes

$$\bar{J}_{\bar{\pi}} = \frac{1}{N}\sum_{i=1}^{N}\sum_{t=0}^{\infty}\gamma^t \cdot \mathbb{E}\left[R(\bar{s}_t^i, a_t^i)|\bar{s}_0, \bar{\pi}\right], \tag{5.3}$$

and the optimization problem is changed as

$$\bar{\pi}^* = \arg\max_{\bar{\pi}} \bar{J}_{\bar{\pi}}.$$

What we hope to address is therefore answer if the introduction of the extra state information o_t^i could be beneficial and lead to

$$\bar{J}_{\bar{\pi}^*} > J_{\pi^*}.$$

If the results hold, then we may develop a set of new multi-agent interactive learning methods for the single agent. Please note that o_t^i can be raw measurement information or encoding features from observation of other agents.

5.2.2 Competitive Framework for Interaction Learning

We propose a framework to exploit competitive information among multiple agents, shown in Fig. 5.2. The main purpose is to acquire knowledge from comparative data. We run a multi-agent competitive task to obtain additional comparable information and then distinguish positive or negative data for better training. It is akin to the contrasting representation between anchor/positive/negative samples. Using the `Walker2d` scenario as an illustration, shown in Fig. 5.3, if a walker demonstrates exceptional performance, it must be the fastest participant in the race. Concurrently, the agent gathers distinctive relative details, like maintaining a consistently positive relative speed compared to others. This situation creates an auxiliary signal connecting high-speed running and favorable data (comprising state-action-reward pairs).

The framework includes mainly two parts: Training and Evaluation. We train agents with competitive observation during the training stage, while only proprioceptive state is required during the evaluation stage. The policy training part conducts competitive tasks among multiple agents: Agents collect full information involving proprioceptive state and competitive observation. The competitive information vector is a concatenation of relative observation against rivals.

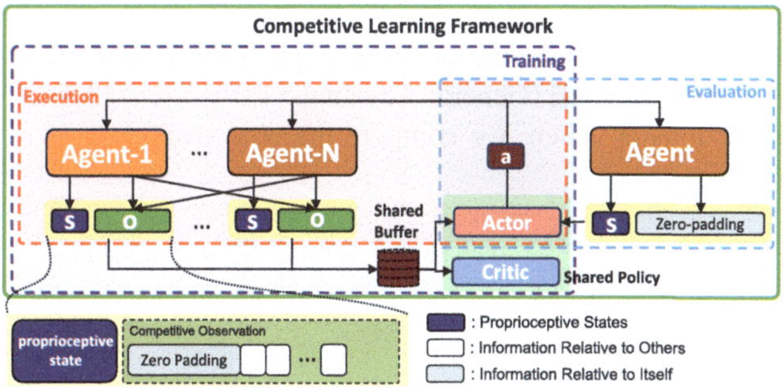

Fig. 5.2 Framework for learning knowledge from interactive information. Where a denotes actions, s represents the proprioceptive state, and o is the competitive observation. (Reproduced with permission from [22])

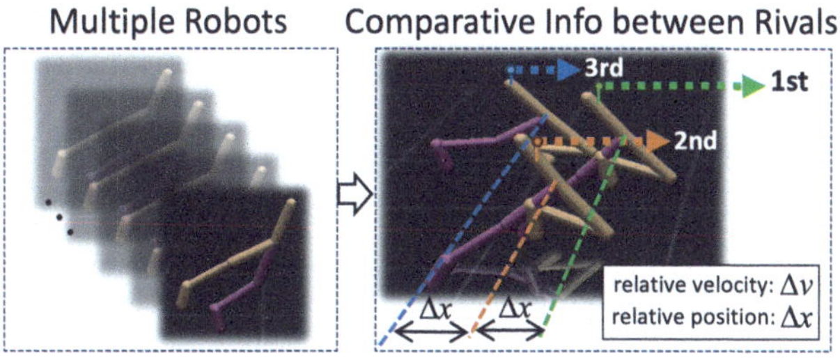

Fig. 5.3 The proposed Self-Interest Competition Environments **Multi-agent-Race**. Three walker2d agents racing is illustrated as an example. (Reproduced with permission from [22])

We use the classical actor-critic algorithm PPO and its multi-agent variant [48] as the algorithmic benchmark because PPO has shown generally favorable results in continuous action control tasks. It becomes normal single-agent learning when there is only one agent in the race.

Since agents are completely homogeneous and exclusively self-interested, we adopt a shared policy among agents. During the training stage, data are sampled distributedly, but the policy is trained centrally since we only maintain one shared policy (actor) $\pi\left(a^i \mid \bar{s}^i; \boldsymbol{\theta}\right)$ for all agents, and one shared value (critic) $v(\bar{s}^i; \boldsymbol{\phi})$ to approximate value function $V_{\boldsymbol{\phi}}$, where $\boldsymbol{\theta}$ and $\boldsymbol{\phi}$ denote parameters of the shared policy/value network. Simultaneously, the experience replay buffer $\mathcal{D}$ is also shared across agents to aggregate all experience $\tau = \{\bar{s}, a, r\}$, and $r = R(\bar{s}, a)$ is the reward given to agents after taking action a under newly designed state $\bar{s}$. In the training part, critic learns an optimal $\boldsymbol{\phi}^*$:

$$\boldsymbol{\phi}^* = \underset{\boldsymbol{\phi}}{\operatorname{argmin}} \frac{1}{\|\mathcal{D}\|\, T} \sum_{\tau \in \mathcal{D}} \sum_{t=0}^{T} \left(V_{\boldsymbol{\phi}}(\bar{s}_t) - R(\bar{s}_t, a_t)\right)^2 . \tag{5.4}$$

Here the competitive observation o is introduced as an additional auxiliary signal where we have $\bar{s} = [s, o]$, helping the critic to get a more accurate estimation of experience. We also use GAE (General Advantage Estimation) [37] to estimate the advantages of actions $\hat{A}$. Then, we compute the objective function to update the policy:

$$
\begin{aligned}
& L(\bar{s}, a, \bar{\alpha}, \theta^{new}, \theta^{old}) = \\
& \min\Big(\frac{\pi_{\theta^{new}}(a \mid \bar{s})}{\pi_{\theta^{old}}(a \mid \bar{s})}\hat{A}^{\pi_{\theta^{old}}}(\bar{s}, \bar{\alpha}), \\
& \text{clip}\Big(\frac{\pi_{\theta^{new}}(a \mid \bar{s})}{\pi_{\theta^{old}}(a \mid \bar{s})}, 1-\epsilon, 1+\epsilon\Big)\hat{A}^{\pi_{\theta^{old}}}(\bar{s}, \bar{\alpha})\Big).
\end{aligned}
\tag{5.5}
$$

We define the probability ratio between the shared new policy and the old one as $\frac{\pi_{\theta^{new}}(a|\bar{s})}{\pi_{\theta^{old}}(a|\bar{s})}$. $\bar{\alpha}$ is the action set of all agents, excluding the current agent. Hyper-parameter ϵ is to limit the difference between old and new policies within a small range.

By sharing experience buffer $\mathcal{D}$, we attain a contrasting effect from competitive data during the critic training. This strengthens the association between rewards and positive/negative actions. The process described by (5.4) is similar to the step of labeling positive and negative samples for training in CL [8]. The difference lies in the fact that we utilize reinforcement learning reward measurement that naturally provides label-like signals to label competitive messages. Through incorporating the contrast provided by the multi-agent competition, competitive representation facilitates a clearer and more explicit understanding of the relationship between state and reward.

As illustrated at the bottom of Fig. 5.2, we consider the new state of the i-th agent $\bar{s}^i = [s^i, o^i]$ to be a concatenation of proprioceptive state s^i and competitive observation o^i, or we call it competitive information which is obtained by comparing with other participants. The first part refers to the sensory feedback that agent receives from joints and muscles, providing measurements of motions and torques. The second part encompasses relative information that agents treat as competitive pressure. We denote $o^i = [o^{i1}, o^{i2}, ..., o^{ij}, ..., o^{iN}]$, where o^{ij} is the observation of the agent i regarding to agent j. Competitive observation can be any differentiable signals or any other comparable features. In this chapter, competitive information of the i-th agent is a concatenation of differences in velocities and displacements $o^{ij} = [x^j - x^i, v^j - v^i]$, which is related to all agents, including itself. x^i, x^j and v^i, v^j represent displacement and velocity of the i-th and j-th agents, respectively. Thus, the length of new state $\bar{s}$ depends on the number of participants in one race.

During the evaluation part, competitive messages are unnecessary, and agents exclusively rely on the proprioceptive state. To maintain alignment with the input dimensions of the neural network, we do zero padding on the missing dimensions.

5.2.3 *Empirical Results*

We extend the single-agent running to multi-agent racing and conduct experiments on four types of agents: `Ant`, `Hopper`, `Walker2d`, `HalfCheetah`.

We primarily consider several benchmarks for comparison to explore the effectiveness of our approach. $\boldsymbol{\theta}$ and $\boldsymbol{\phi}$ are used to represent shared parameters, and then θ^i and ϕ^i denote separated parameters. X represents the agent number of the multi-agent system. Our experiments are conducted in 2, 3, 4, and 5 agent races, respectively, where $X = 2, 3, 4, 5$, to show the effect of different scales of competition.

***S*A** PPO training in a single-agent environment. We have $\pi(a \mid s; \boldsymbol{\theta})$ and $v(s; \boldsymbol{\phi})$.

***X*A-Sh-Decent** Training in X number of agents environment but with no competitive information. Here we define the knowledge-level information sharing as agents sharing the same networks and experiences. The agent uses its decentralized local state as the critic input where we have $\pi\left(a^i \mid s^i; \boldsymbol{\theta}\right)$ and $v(s^i; \boldsymbol{\phi})$.

***X*A-Sh-Cent** Training in X number of agents environment but without competitive information. Policy and experience are shared. Agents use a centralized global state as critic input. We have $\pi\left(a^i \mid s^i; \boldsymbol{\theta}\right)$, and $v(s; \boldsymbol{\phi})$, in which s is the global observation, denotes the concatenation of $(s^1, ..., s^N)$.

***X*A-Sp-Decent-Comp** Training in X number of agents environment with competition. Every agent trains its own specific policy using local observation as value input. We can simply consider it to be a parallel training of SA with competitive observation. It is different from the proposed approach: Although competitive information is applied to agents, no contrastive representation is formed because experiences are not shared. Therefore, the agent cannot improve itself by learning from others, where we have $\pi\left(a^i \mid \bar{s}^i; \theta^i\right)$ and $v(\bar{s}^i; \phi^i)$.

***X*A-Sh-Decent-Noi** Training in X number of agents environment with random noise. Previous studies argued that noise has a significant performance improvement on multi-agent learning rather than global information [18]. To validate that agents indeed acquire effective experience from competitive information, we replace with zero-mean random noise as a control group where we have $\pi\left(a^i \mid [s^i, n]; \boldsymbol{\theta}\right)$ and $v([s^i, n]; \boldsymbol{\phi})$, in which n denotes the noise.

***X*A-Sh-Decent-Comp (Proposed)** Training in X number of agents environment with competitive input. Agents share policy and experience where we have $\pi\left(a^i \mid \bar{s}^i; \boldsymbol{\theta}\right)$ and $v(\bar{s}^i; \boldsymbol{\phi})$.

5.2.3.1 Role of Competitive Interaction Information

We benchmark the proposed approach against baselines using three-agent environments as a representative case. It is important to note that our method might

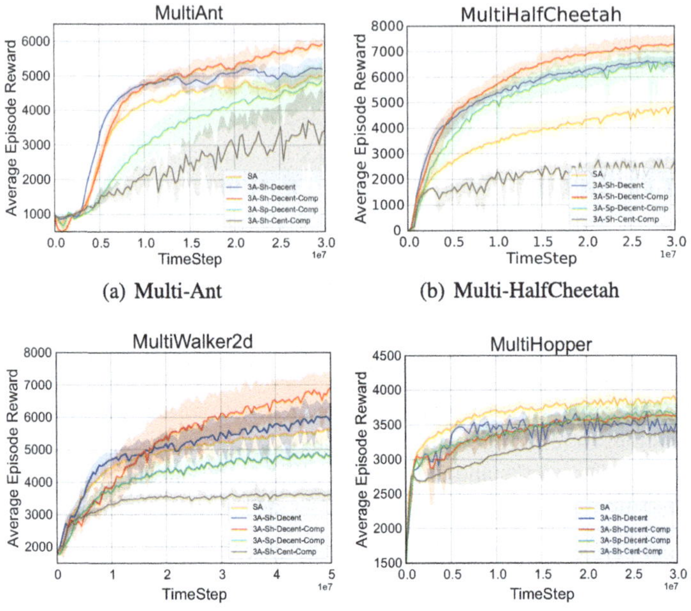

(a) Multi-Ant (b) Multi-HalfCheetah

(c) Multi-Walker2d (d) Multi-Hopper

Fig. 5.4 Performance of different settings on three-agent environments. SA: Yellow. 3A-Sh-Decent: Blue. 3A-Sh-Decent-Comp: Red. 3A-Sp-Decent-Comp: Green. 3A-Sh-Cent-Comp: Brown. (Reproduced with permission from [22])

yield better performance in environments with a different number of agents. Results are illustrated in Fig. 5.4. Our proposed approach, training with competition and shared policy (3A-Sh-Decent-Comp), outperforms all baselines on task `MultiAnt`, `MultiHalfCheetah`, and `MultiWalker2d`. Besides, we notice that SA and 3A-Sh-Decent have a similar trend, but the latter slightly exceeds the former. This occurs because multiple agents gather more data leading to a more balanced distribution of data abundance and variance and increasing accuracy in the sampling results. Comparing 3A-Sh-Decent and 3A-Sh-Decent-Comp, we can conclude that agents learn valuable knowledge from additional comparative information. Especially in task `MultiAnt`, `MultiHalfCheetah`, and `MultiWalker2d`, we observe around 13%, 11%, and 16% improvement, respectively.

However, it does not work on `MultiHopper`. In contrast, settings of competitive training seem to negatively affect the results. This phenomenon could be attributed to the `Hopper` task being overly simplistic, and the competitive observation might drown out the proprioceptive state. On the other hand, it appears that the dynamics of the `Hopper` have been thoroughly explored and understood, leading to little improvement under competition.

5.2.3.2 Ablation Studies

We perform a series of ablation studies to obtain a more profound understanding of how competitive scenarios yield valuable data for the learning process.

Policy and Buffer Sharing We conducted controlled experiments on four environments depending on whether the policy network and experience reply buffer are shared or not, using the three-agent racing game as an example. In 3A-Sp-Decent-Comp, agents enjoy independent policies, while agents share one policy and buffer in 3A-Sh-Decent-Comp. Results are shown in Fig. 5.4. Irrespective of the varying tasks, agents with a shared network consistently demonstrate superior performance compared to their counterparts utilizing independent policy. This substantiates the explanation of (5.4) discussed above.

Moreover, as there is no interaction and cooperation between competitors, the value network need not coordinate the centralized global states of all agents. By contrast, centralized value inputs might confuse agents and harm the performance of evaluation, illustrated as brown lines in Fig. 5.4.

Learning from Competition There remains ongoing debate regarding the influence of external information on the effectiveness of training. Some studies have proved that adding more comprehensive information to the MARL task as additional states could improve the performance [10, 47]. However, others believe the network might treat external messages as noise that encourages more exploration [18], rather than learning valuable features from messages.

We employ random noise that is equivalent in length to the competitive information message as a controlled group, named XA-Sh-Decent-Noi. 3S-Sh-Decent is also taken as the comparison using fixed zero padding instead. We compare the performance of 3A-Sh-Decent, 3A-Sh-Decent-Noi, and 3A-Sh-Decent-Comp in Fig. 5.5. Our results can corroborate the viewpoints presented in [18]: Assigning

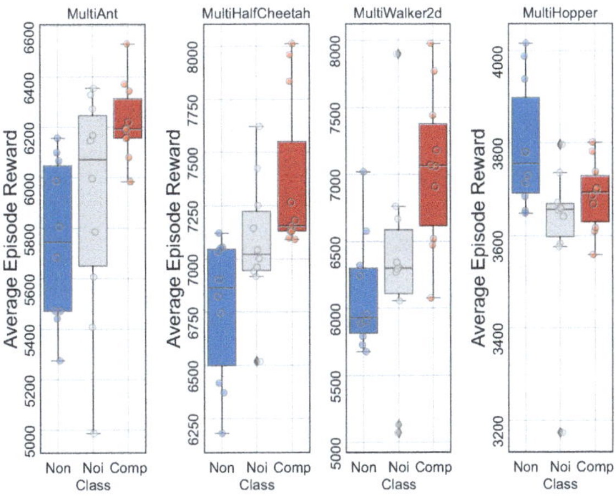

Fig. 5.5 Comparison between 3A-Sh-Decent (**Non**: non-competitive information as inputs), 3A-Sh-Decent-Noi (**Noi**: noise information as inputs), and 3A-Sh-Decent-Comp (**Comp**: competitive information as inputs). (Reproduced with permission from [22])

noisy observations as inputs enhances the network's exploration capability, consequently leading to higher rewards. Simultaneously, we also prove that agents can acquire knowledge from competitive observation, leading to a higher reward surpassing those of the noise experiments.

5.2.4 Discussion

We find there exists an unavoidable correlation between the results of games, the complexity of agents, and the number of racers involved. To explore how the number of racers affects the results, we conduct experiments with different numbers of participants, shown in Fig. 5.6.

We compare the outcomes of utilizing competitive information in different environments and varying numbers of runners in tasks. For the sake of fairness, we record data after convergence. The results shown in Fig. 5.6 indicate that training with competitive pressure can promote performance regardless of racer number, except `Hopper`, because we believe its potential has been fully explored. Our proposed method makes a great improvement on `Walker2d` and `Ant`, and then a slight improvement on `HalfCheetah` task.

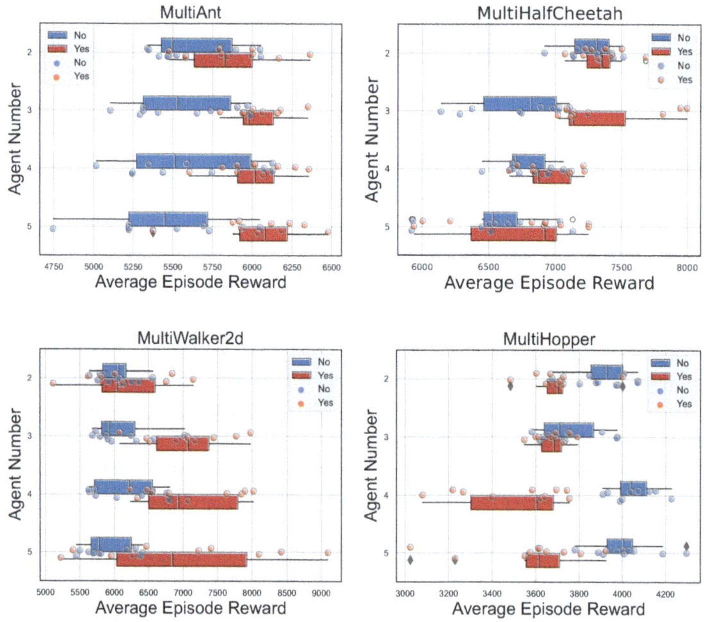

Fig. 5.6 Results on various numbers of agents with/without competitive information. (Reproduced with permission from [22])

Furthermore, we discovered that on more challenging tasks like more complex agents, the advantages of utilizing competition are more pronounced. For example, agent `Ant` simulates in the three-dimensional environment which is more complex, and agents achieve higher running speeds with an increase in the number of competitors. However, for agents with simpler structures, indiscriminately increasing the number of competitors could potentially lead to a decrease on final rewards. Declines are observed on 2D agents `Hopper`, `HalfCheetah`, and `Walker2d`. This is because agents with simple structures and straightforward dynamics can easily reach their ability boundary, while agents with complex structures and challenging tasks require more sophisticated controllers. We obtain the best performance on 2-`HalfCheetah`, 3-`Walker2d`, 5-`Ant`, and single-`Hopper`.

In addition, an excessive number of competitors, however, introduce an excessive amount of competitive information, resulting in an increase in the observation dimension. The additional signals representing competition could overshadow the proprioceptive signals of the agent. This leads to challenges in agent learning when the number of runners becomes excessive.

Although our work primarily focuses on PPO as the core algorithm, our framework is adaptable to any other on-policy multi-agent algorithms. On-policy algorithms, by not relying on past experiences, can establish precise comparative features within a single batch of data sampled by the current policy. The framework cannot adapt to off-policy algorithms. The preservation of old experiences is common practice in off-policy methods, leading to the absence of a baseline value for evaluating the comparison of information along a long-term training process.

5.3 Adversarial Skill Learning for Robust Manipulation

In recent years, policies trained by reinforcement learning (RL) methods [4, 14] have been widely used in robot control and have achieved great success in lots of scenarios. However, ordinary RL policies are usually fragile and can easily fail when disturbances occur. In a real-world environment, both internal and external disturbances inevitably exist. Thus, it is crucial for a practical robot to be robust enough to adapt to them automatically.

In fact, robust control of robots has been a heated research topic for decades [6, 32, 41]. With the application of reinforcement learning methods in robot control, researchers have developed several methods to improve the robustness of reinforcement learning [30, 34]. Among them, adversarial reinforcement learning is a new idea developed in recent years [12, 13, 33, 35, 45].

To the best of our knowledge, few researchers have investigated different types of disturbance in robotic manipulation tasks comprehensively and proposed a robust RL method, which is more generalized. From our perspective, the disturbances can be divided into two classes: internal disturbance and external disturbance. We define the internal disturbance as the sum of various internal noise on the joints of a robot, making the robot unable to carry out the ideal action given by its controller. The

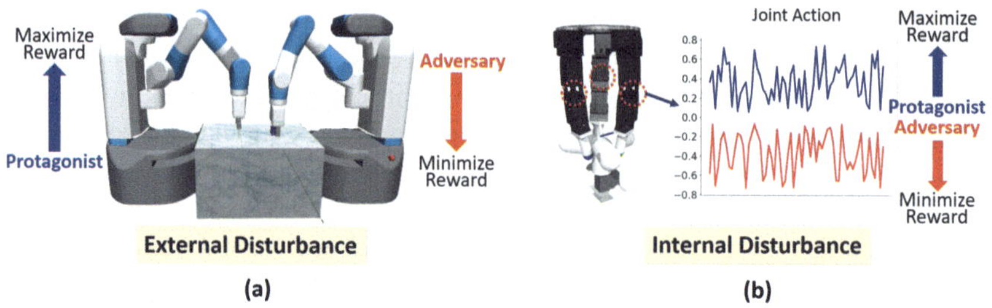

Fig. 5.7 Adversarial skill learning for robust manipulation. (Reproduced with permission from [23])

external disturbance is executed outside the robot and can hinder the robot from accomplishing its task. It is important to analyze these two kinds of disturbance separately and develop a general method to improve the robot's robustness against both of them. The main idea of this chapter is demonstrated in Fig. 5.7. For the external and internal disturbance in robotic manipulation tasks, protagonist agents try to maximize the reward, while adversary agents try to minimize it. The protagonist agent is expected to be more robust after the adversarial training.

In this chapter, we develop an adversarial training method based on soft actor-critic (SAC) [17], which can improve robot robustness against internal and external disturbances on robotic manipulation tasks. Our proposed algorithm can be applied against both the internal noise in robot joint space and the external disturbance from another adversarial robot or human being.

5.3.1 *Background and Preliminary*

Before we illustrate our algorithm in detail, we need to make some necessary explanations for the involved basic concepts and preliminaries.

5.3.1.1 Standard Reinforcement Learning on MDPs

In this chapter, we consider a standard finite-horizon, discounted, Markov Decision Process (MDP) defined by a tuple $< \mathcal{S}, \mathcal{A}, r, \mathcal{P}, s_0, \gamma, T >$, where $\mathcal{S}$ and $\mathcal{A}$ are the sets of feasible state and action; $r(s, a) : \mathcal{S} \times \mathcal{A} \rightarrow \mathbb{R}$ denotes the reward function on state s and action a; $\mathcal{P}(s'|s, a) : \mathcal{S} \times \mathcal{A} \times \mathcal{S} \rightarrow [0, 1]$ characterizes the dynamics of the environment and defines the transition probability to next-step state s' if the agent takes action a at current state s; s_0 is the distribution of initial state, T is the maximum finite-horizon time, and $\gamma \in (0, 1)$ is the discount factor. An agent policy $\pi(a|s) : \mathcal{S} \times \mathcal{A} \rightarrow [0, 1]$ defines the probability of choosing action a at state s. The goal of a given MDP is to find a policy π_μ parameterized by μ that maximizes the expected cumulative discounted reward:

$$J^{\mu} = E_{\pi_{\mu}}[\sum_{t=0}^{T} \gamma^{t} r(s_t, a_t)|s_0 = s], \tag{5.6}$$

where the actions a in the MDP are sampled via $a \sim \pi_{\mu}(a|s)$.

5.3.1.2 Soft Actor-Critic

Soft actor-critic (SAC) is an algorithm that optimizes a stochastic policy in an off-policy way, forming a bridge between stochastic policy optimization and DDPG-style [24] approaches.

SAC is an actor-critic technique that consists of two models: Actor and Critic. The critic network is the function approximation of action-state value function $Q(s, a)$ which evaluates the quality of state-action pair (s, a). It uses the prior concepts of experience replay and target network for its update. The actor policy is trained to maximize a trade-off between expected cumulative discounted reward and policy entropy. In SAC, we consider an entropy augmented objective:

$$J^{\mu} = E_{\pi_{\mu}(a|s)}[\sum_{t=0}^{T} \gamma^{t} r(s_t, a_t) + \alpha H(\pi_{\mu}(a|s))], \tag{5.7}$$

where the $H(\pi_{\mu}(a|s))$ is the entropy of actor policy $\pi_{\mu}(a|s)$, and α is a temperature hyper-parameter. Optimizing this objective function can bring maximum entropy of the policy to encourage exploration, which can accelerate learning later on. It can also prevent the policy from prematurely converging to a bad local optimum.

5.3.1.3 Sparse Reward

Most of the robot manipulation tasks are sparse reward reinforcement learning problems. The agent can only get a reward when it accomplishes the task. Since completing the task by random exploration is of extremely low probability, it is hard for the agent to explore and get successful experiences. To overcome the exploration difficulty in robotic manipulation tasks, we adopt a data augmentation method—Hindsight Experience Replay (HER). The key idea of HER [3] is to relabel failed rollouts as successful ones. During the exploration process of a sparse reward task, the agent fails in most rollouts. To make full use of these failure data, we relabel the final state of a failed trajectory as the desired goal state so that the agent can get the same reward when it reaches this relabeled state as the original goal. Both original successful transitions (s_t, g, a_t, r) and the relabeled transitions (s_t, g', a_t, r') are pushed into the replay buffer. Then, we sample from the replay buffer to optimize the agent policy.

5.3.2 *Adversarial Skill Learning*

Here we outline the adversarial skill learning in an environment with disturbance. We first introduce the minmax adversarial training mechanism to model the environment with disturbance. Through two different forms of disturbance, we then sort out their application scenes, respectively. Finally, we provide a practical implementation for our proposed method from the stochastic policy.

5.3.2.1 Formulating Adversarial Attack

The adversarial training setting can be expressed as a two-player γ discounted zero-sum Markov game. In this chapter, we consider a robust MDP [42] as the tuple: $(\mathcal{S}, \hat{\mathcal{A}}, \bar{\mathcal{A}}, \mathcal{P}, r, \gamma, s_0)$, where $\hat{\mathcal{A}}$ and $\bar{\mathcal{A}}$ are the continuous actions the players can take. $\mathcal{P} : \mathcal{S} \times \hat{\mathcal{A}} \times \bar{\mathcal{A}} \times \mathcal{S} \to \mathbb{R}$ is the transition probability density, and $r : \mathcal{S} \times \hat{\mathcal{A}} \times \bar{\mathcal{A}} \to R$ is the reward of both players. We parameterize actor policy π_θ and adversary policy π_ϕ as θ and ϕ, respectively. The optimization problem is formulated as

$$\min_{\phi} \max_{\theta} E[\sum_{t=0}^{T} \gamma^t r(s_t, \hat{a}_t, \bar{a}_t)|\pi_\theta, \pi_\phi], \tag{5.8}$$

where $\hat{a}_t$ and $\bar{a}_t$ are sampled from π_θ and π_ϕ.

A zero-sum two-player game can be seen as the protagonist maximizing the γ discounted reward, while the adversary minimizing it [25, 31, 34]. The solution to a zero-sum game is a minimax search; it is a result of Nash equilibrium. In the zero-sum game setting, we call actor policy π_θ of protagonist agent that we wish to be more robust *protagonist* and the actor policy π_ϕ of adversary agent who tries to fail the task *adversary*. And the adversary is used to simulate the possible disturbance in the environment. Adversarial training allows the protagonist to encounter more severe situations, especially those dangerous but unexplored situations, thereby making it more adaptable to noisy environments. It also enables the adversary to learn to hinder the protagonist more effectively. Both the protagonist and the adversary receive feedback from the environment, and they both give an action based on the observation. The difference is that the adversary makes an action that tries to fail the task and makes the protagonist get a lower reward [13]. To simulate the real-world situation that the protagonist is disturbed by internal noise and external disturbance, we utilize two types of disturbances.

Internal Robot Joint Disturbance Considering the actual robot with motor noise, the noisy action can be added into the robot joint space, which often happens when the joint motors of a robot are in bad condition. Formally, we will construct the noisy action as

$$a = \hat{a} + \alpha \cdot \bar{a}, \tag{5.9}$$

where $\hat{a}$ and $\bar{a}$ are the protagonist and adversary action of which each dimension is within the range $[-1, 1]$, and α is the parameter to control the maximum possible amplitude of the disturbance. For convenience, we will call α the amplitude parameter in the following text and figures.

External Robot Disturbance We use another robot acting to hinder the protagonist robot during the task, which is a perfect simulation when some human being or robot is really trying to disturb the protagonist robot and fail the task. Formally, we will construct the noisy action as

$$a = ||[\hat{a}, \bar{a}], \tag{5.10}$$

where $||$ represents concatenation, $\hat{a}$ is our protagonist robot action, and $\bar{a}$ is our external adversarial robot action. We do not use a similar α in a) to constrain the $\bar{a}$ here, because we think it makes sense that the maximum possible amplitude of the disturbance on robot joint space can vary in reality, but it is not that reasonable to modify the action amplitude of another robot.

5.3.2.2 Adversary Optimization

We further analyze the practical optimization process for our adversarial training for the stochastic policies built on the soft actor-critic framework.

In the case of stochastic policy, we extend the standard soft actor-critic approach to our adversarial version, called *Adv-SAC*. Here we note the protagonist agent's actor and critic networks as Q_p and μ_p and use Q'_p and μ'_p to represent its target actor and critic networks. For the adversary agent, Q_a, μ_a, Q'_a, μ'_a represent the adversary actor, critic networks, and corresponding target actor and critic networks.

As an entropy-regularized reinforcement learning, the critic objectives for both protagonist and adversary agents can be formulated as

$$L(\theta_{Q_p}) = \frac{1}{N}\sum_i (y_i^p - Q_p(s_i, a_i|\theta^{Q_p}))^2, \tag{5.11}$$

$$L(\theta_{Q_a}) = \frac{1}{N}\sum_i (y_i^a - Q_a(s_i, a_i|\theta^{Q_a}))^2, \tag{5.12}$$

where the protagonist target value and the adversary target value are

$$y_i^p = r_i + \gamma Q'_p(s_{i+1}, \mu_p(\cdot|s_{i+1}) - \alpha \log \mu_p(\cdot|s_{i+1})), \tag{5.13}$$

$$y_i^a = r_i + \gamma Q'_a(s_{i+1}, \mu_a(\cdot|s_{i+1}) - \alpha \log \mu_a(\cdot|s_{i+1})). \tag{5.14}$$

We can note that the adversary agent critic network is the same as the protagonist where we just replace Q_p Q'_p and μ_p in Eq. 5.11 with Q_a Q'_a and μ_a.

The main difference is from the policy optimization part. For the protagonist policy, our objective is to maximize the expected reward with maximizing entropy-regularized term.

$$\begin{aligned}\nabla_{\theta^{\mu_p}} J(\theta) =& \nabla_a Q_p(s,a) \nabla_{\theta^{\mu_p}} \mu_p(s|\theta^{\mu_p}) \\ &+ \nabla_\theta \log \pi_\theta^{\mu_p}(a|s) Q_p^H(s,a).\end{aligned} \tag{5.15}$$

For the adversary policy, our objective is to minimize the expected reward with maximizing entropy-regularized term.

$$\begin{aligned}\nabla_{\theta^{\mu_a}} J(\theta) =& -\nabla_a Q_a(s,a) \nabla_{\theta^{\mu_a}} \mu_a(s|\theta^{\mu_a}) \\ &+ \nabla_\theta \log \pi_\theta^{\mu_a}(a|s) Q_a^H(s,a).\end{aligned} \tag{5.16}$$

The overall process is shown in Fig. 5.8, and the pseudocode is summarized in Algorithm 5.1.

5.3.3 Experiments

For the experiments below, we investigate the following questions:

1. Under the disturbance from internal robot joints and external environment, is the protagonist manipulation policy robust enough to accomplish the given task?
2. Is our adversarial attack a more aggressive disturbance than random noise in the robot manipulation tasks?

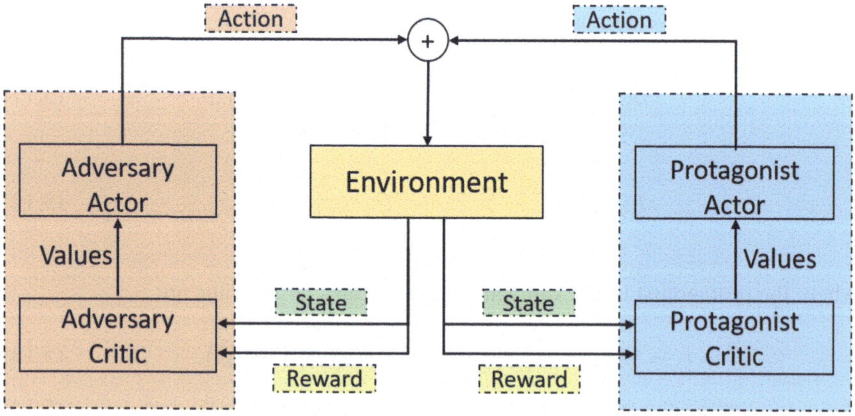

Fig. 5.8 Adversarial attack mechanism. (Reproduced with permission from [23])

Algorithm 5.1 Adversarial training

Require: Randomly initialize deterministic protagonist actor μ_p and adversary actor μ_a with are parameterized as θ^{μ_p}, θ^{μ_a}, randomly initialize protagonist critic network $Q_\theta(s, \hat{a})$ and adversary critic network $Q_\phi(s, \bar{a})$.
for iteration $i = 1$ to N_{iter} **do**
  \\ update protagonist agent policy $\theta_i^{\mu_p} \leftarrow \theta_{i-1}^{\mu_p}$.
  for episode=1 to M **do**
    Sample $\hat{a}, \bar{a}$ from $\theta_{i-1}^{\mu_p}, \theta_{i-1}^{\mu_a}$. And execute action $a = \hat{a} + \alpha\bar{a}$ or $a = |||\hat{a},\ \bar{a}]$
    Observe reward r and next state s' from state s.
    Relabel data $(s, \hat{a}, r, s')$ via HER.
    Update protagonist critic $Q_\theta(s, \hat{a})$from Eq. (5.11).
    Update protagonist actor θ^{μ_p} from Eq. (5.15).
  end for
  \\ update adversary agent policy $\theta_i^{\mu_a} \leftarrow \theta_{i-1}^{\mu_a}$.
  for episode=1 to N **do**
    Sample $\hat{a}, \bar{a}$ from $\theta_i^{\mu_p}, \theta_{i-1}^{\mu_a}$. And execute action $a = \hat{a} + \alpha\bar{a}$ or $a = |||\hat{a},\ \bar{a}]$
    Observe reward r and next state s' from state s.
    Relabel data $(s, \bar{a}, r, s')$ by HER.
    Update adversary critic $Q_\phi(s, \bar{a})$ from Eq. (5.12).
    Update adversary actor θ^{μ_a} from Eq. (5.16).
  end for
end for
return $\theta_{N_{iter}}^{\mu_p}, \theta_{N_{iter}}^{\mu_a}$

3. Can the results of the real-world experiment support a similar conclusion with the simulation experiment results?

We conduct (1) robot claw turning task *DClawTurnFixed* on ROBEL platform [1] in simulation environment and real-world environment; (2) the *DoubleArmPick* task custom in the gym [7] environment. In the first task, we add internal disturbance on Robel robot joints. In the second task, we use another adversarial robot to execute external disturbance on the protagonist robot.

The *DClawTurnFixed* task in the ROBEL environment requires the robot claw(D'Claw) to turn the triple-arm object over 180°. The triple-arm object is initialized at a fixed angle −180°. If the robot claw turns the object to above 0° once within 40 timesteps, it succeeds in this game. As is shown in Fig. 5.9b, in real-world experiments, we use orange markers on the object and the ground to mark the object angle and the target angle. In Fig. 5.11, we use a blue line to show the object angle clearer. The *DoubleArmPick* task is a double robot version of the *FetchPickAndPick* task in the OpenAI gym [7]. It requires a robot to pick the block on the table and place it at the goal position in the air or on the table within 50 timesteps. Both block and goal positions are initialized randomly. The first and second tasks are set in dense and sparse reward modes separately, indicating our method is suitable for both situations.

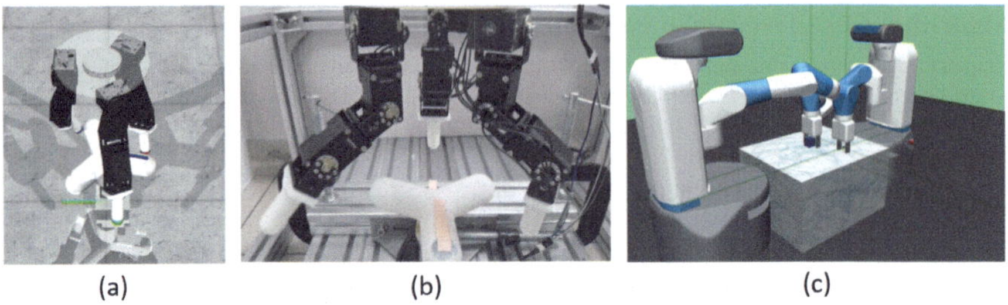

Fig. 5.9 (**a**) *DClawTurnFixed* task in ROBEL [1] simulation environment. (**b**) Real-world *DClaw-TurnFixed* platform using ROBEL [1] D'Claw robot. (**c**) *DoubleArmPick* task custom in gym [7] environment. (Reproduced with permission from [23])

5.3.3.1 Internal Disturbance on Robel Robot

In the Robel robot experiments, we simulate the situation when the joint motors of a robot are in bad condition and have noise on them. We hope to figure out how the noise on the joint motors will influence the robot performance and develop a robust policy for robots to overcome the disturbance on joint space. In this case, the ideal robot action on joint space is represented as $\hat{a}$. The disturbance on joint space is represented as $\alpha\bar{a}$, where α stands for the maximum possible amplitude of the disturbance, and $\bar{a}$ represents its character, which can be random noise or the output action of an adversarial agent. Each dimension of $\hat{a}$ and $\bar{a}$ is within the range $[-1, 1]$. The random noise obeys the uniform distribution on $[-1, 1]$. So the real action on the robot joints can be given by $a = \hat{a} + \alpha\bar{a}$, which means the real action on each joint is the linear combination of the ideal action and a disturbance with a maximum possible amplitude α. We will call α amplitude in short in the following text and figures.

Simulation Experiment As is shown in Fig. 5.9a,b, a Robel D'Claw robot is trained to accomplish the *DClawTurnFixed* task.

Utilizing the adversarial SAC training method, a protagonist policy and an adversary policy are trained alternately, and each is trained for 20,000 episodes. A normal SAC policy is also trained for 20,000 episodes to be tested as a contrast baseline. We choose $\alpha = 0.6$ during the adversarial training process because after lots of experiments, we find that this α value can help train the most robust protagonist and the most aggressive adversary in this task. Then, we test the trained policies in a wide range of α values to test the adaptability of the trained policies.

We use the output action $\bar{a}$ of the adversary to attack the protagonist and the normal SAC agent. α varies from 0.0 to 1.4, and the gained rewards of both agents are shown in Fig. 5.10a. When α is below 0.2, both agents can accomplish the task and get a high reward. When α is between 0.2 and 0.6, our protagonist can still accomplish the task perfectly and maintain the high reward, while the normal SAC agent fails and gets a meager reward. When α rises to above 0.8, both agents fail.

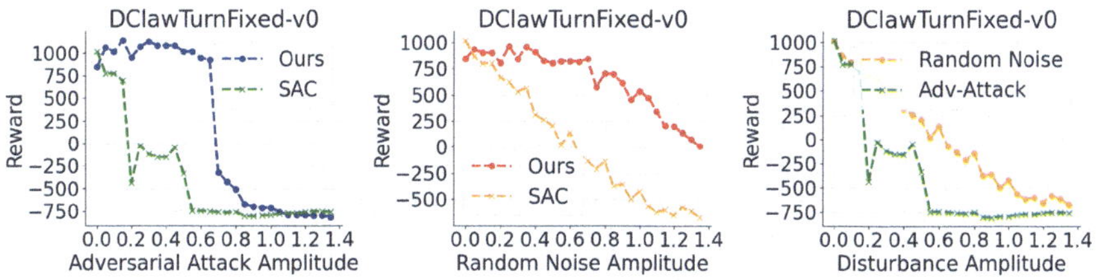

Fig. 5.10 (**a**) Compared with normal SAC agent, our protagonist gets obviously higher reward against adversarial attack on joint space. (**b**) Compared with normal SAC agent, our protagonist gets higher reward against random noise on joint space. (**c**) Attacking the same normal SAC agent on its joint space, the adversary decreases the reward more dramatically than random noise. (Reproduced with permission from [23])

Figure 5.10b suggests that our protagonist has higher robustness than the normal SAC agent against the random noise on joint space. When the amplitude α of the random noise rises, the rewards of both agents will decline, but the protagonist can keep getting a higher reward than the SAC agent in a large range of α, indicating that the protagonist accomplishes the task better in the same noisy condition.

The effects of adversarial attack and random noise on the normal SAC agent are compared in Fig. 5.10c. At any same amplitude α below 1.4, the reward under adversarial attack is significantly lower than that under random noise, showing that adversarial attack is smarter and more aggressive than the random noise to hinder the protagonist and fail the task.

To sum up, the simulation experiments on the Robel robot show that the trained protagonist can obtain significantly better robustness against the adversarial attack and random noise. Moreover, the trained adversary appears to be far more aggressive than a random attack in all these experiments.

Real-World Experiment We also conducted real-world experiments on the Robel D'Claw robot in the same task. As is shown in Fig. 5.11, the D'Claw robot is hung on a frame, and the triple-arm object with a motor on its base is fixed on the floor to record the angle of the object during the experiments and reset the object to the $-180°$ initial position before each game.

Since the simulation environment and our real-world experiment environment are basically consistent, we apply the protagonist policy and the adversarial policy trained in the simulation environment directly on the real robot. Also, a normal SAC policy is trained in the simulation environment and used on the real robot for contrast. Figure 5.11 shows the situation when the amplitudes of adversarial attack or random noise on the joint space are both 0.8. The first three columns of Fig. 5.11 show the process of turning the object. The last column includes the curves of the object angle varying with the timestep. In each game, if the robot claw can turn the object from $-180°$ to over $0°$, it succeeds.

In the first row of Fig. 5.11, when the protagonist is acting against a random noise with amplitude 0.8, it can easily accomplish the task. The curve indicates

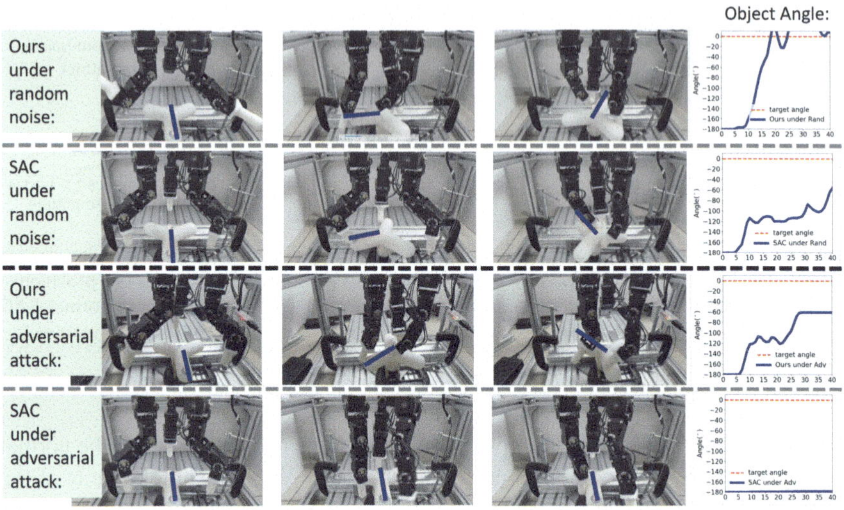

Fig. 5.11 Experiments result in reality

that the object is turned over 180° very quickly, and Table 5.1 shows a success rate around 94%. In contrast, when the normal SAC agent acts against random noise, it sometimes fails to turn the object to an angle large enough to succeed, getting a success rate of around 82%. Its turning process and the object angle curve are recorded in the second row of Fig. 5.11. The lower two rows of Fig. 5.11 compare the performance of the protagonist and normal sac agent against adversarial attack. Our protagonist can turn the object to about 120° but rarely over 180° to success. On the contrary, the normal SAC agent becomes totally out of function and can hardly move the object at all.

Considering that the real-world platform is impossible to be in the exact same condition as the simulation platform, the results of the real-world experiment are slightly different from the results of its simulation experiments on some specific numbers, but the general trends are the same. Both real-world and simulation results indicate that the adversarial training can train a protagonist with higher robustness and an adversary, which is more aggressive than random noise. More details can be seen in the videos.

5.3.3.2 External Disturbance in *DoubleArmPick* Task

Besides the internal disturbance like motor noise, which can be described as $\alpha\bar{a}$ added on the joint space of a robot, there are also some cases when the robot is disturbed by a real person or another real robot. To simulate this kind of situation, we build a double robot environment in gym [7], as shown in Fig. 5.9c. The robot on the right side of the table is the protagonist who tries to accomplish the *DoubleArmPick*

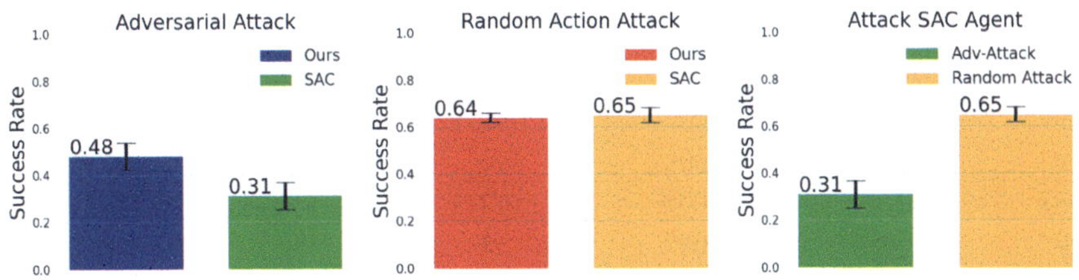

Fig. 5.12 In *DoubleArmPick* task: (**a**) Under the adversarial attack of another robot, our protagonist can get about 20% higher success rate than the normal SAC agent. (**b**) When the adversarial robot takes random action, our protagonist and normal SAC agent have similar performance in success rate. (**c**) Adversarial attack from another robot is far more aggressive than a random attack, causing the normal SAC agent an extra 30% drop in success rate. (Reproduced with permission from [23])

task in sparse reward mode, while the robot on the left side is the adversary who tries to hinder the protagonist and fails the task.

Using our proposed adversarial training method based on SAC, we train a protagonist for the right robot and an adversary for the left robot. Also, we lock the left robot (setting its action to zero in every step) and train a normal SAC agent on the right robot for the *DoubleArmPick* task as a contrast baseline for the following tests.

The protagonist or normal SAC agent on the right is tested in the *DoubleArmPick* environment against the adversary robot on the left. Figure 5.12a shows the experimental results when the left robot executes the trained adversarial policy. Fighting against the adversarial robot, the protagonist can get about 50% of success rate in the *DoubleArmPick* task, while the normal SAC agent can only reach around 30% of success rate, indicating that the adversarial training can improve the robustness of the protagonist agent against the disturbance of another robot.

When the adversarial robot on the left takes a random action instead of the output from an adversarial policy, the adversarial trained protagonist can maintain a performance close to the normal SAC agent. As shown in Fig. 5.12b, it can reach a success rate of above 64%, slightly lower than that of the normal SAC agent.

Comparing the success rate of the normal SAC agent against adversarial action and random action in Fig. 5.12c, it is evident that the trained adversarial policy can take a far more aggressive attack than the random action, causing an extra 30% decline in the success rate of the task. As for real-world experiments on the Robel D'Claw robot, the mean success rate in Table 5.1 of each setting is calculated after 50 times repeated experiments. Under the disturbance of the random noise, the success rate of our method can maintain above 90% within the range of the amplitude from 0.0 to 1.0. In contrast, when the amplitude of the noise rises to 1.0, the success rate of the SAC method declines dramatically to 38%. Under the adversarial attack with an amplitude of 0.5, our method can achieve a success rate of 86%, while the SAC method almost fails every time. Comparing the success

Table 5.1 Success rate of "DClawTurnFixed" experiments

Method	Disturbance	Amp = 0.0	Amp = 0.5	Amp = 0.8	Amp = 1.0
Ours	Rand.	100%	98%	94%	92%
SAC	Rand.	100%	88%	82%	38%
Ours	Adv.	100%	86%	4%	2%
SAC	Adv.	100%	2%	0%	0%

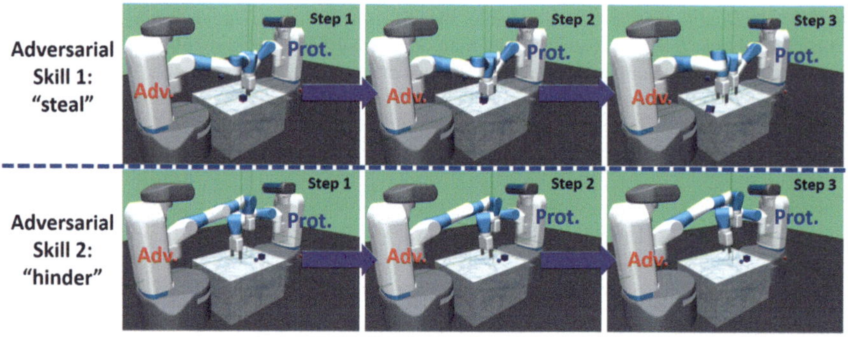

Fig. 5.13 The first row of three figures shows the adversarial robot's process on the left pushes the block off the table to fail the task. (Reproduced with permission from [23])

rates under the adversarial attack and random noise with the same amplitudes, it is obvious that the adversarial attack is more aggressive than the random noise.

Figure 5.13 shows two of the typical skills learned by the adversary to fail the task. In the first row of Fig. 5.13, the block is initialized close to the adversarial robot. The adversary will reach it ahead of the protagonist and then push it off the table, making it impossible for the protagonist to accomplish the task. In the second row of Fig. 5.13, the block is initialized close to the protagonist. In this case, it is hard for the adversary to reach the block first and push it off the table, so it will stretch its arm far ahead and put it in the way of the protagonist robot arm to the block. By doing so, it can hinder the protagonist from reaching the block and thereby fail the task. There are also many other interesting skills learned by the protagonist and the adversary. For example, the two agents will compete to get the block in some cases. Sometimes, the adversary will push away the block that has already been placed in the protagonist's goal position. Sometimes, the protagonist will push away the adversary gripper to get the block. Please find them in the video.

5.3.4 Conclusion

In this chapter, we propose an adversarial training algorithm for robotic manipulation tasks and carry out abundant experiments to test the robustness of the trained protagonist and the disturbance ability of the adversary. This algorithm can improve the robustness of the policy against both internal disturbances in the robot's joint space and external disturbances from other robots. We believe our algorithm can be widely useful in real-world robot applications against inevitable internal and external disturbances. Besides, to the best of our knowledge, this is also the first time that researchers successfully apply the adversarial training method on sparse reward RL tasks. According to the results of both simulation and real-world experiments, our trained protagonist is more robust than normal RL policy, and the trained adversary is obviously more aggressive than random disturbance. More real-world experiments on other robotic manipulation tasks can be a good direction for future work.

5.4 Coevolve Body and Brain in Confrontation

Achieving evolution through cooperation and competition among multiple individuals is pervasive in the biological realm [19, 20, 49], especially on the morphological aspect. For example, athletes usually undergo extensive training to develop and reshape their bodies, along with honing their skills, to achieve better performance in formal competitions. Inspired by this, we redirect our attention to the interactive coevolution of morph and combat tactics within competitions.

Our emphasis is on the strategy that coevolves morph and tactics within two-player games, helping agents to interactively learn fighting skills and fighting designs. Many studies have shown the advantages of reinforcement learning in two-player games like Go and Chess [38, 39]. When considering embodied AI in two-player games, OpenAI provides amazing results that promote the simulation of agents at the control and execution levels [5]. For two-player games where agents have asymmetric morph, meta-learning methods are proposed to solve a long-term continuous adaptation [2]. More embodied learning platforms of multi-agent cooperation and competition are also proposed by DeepMind [16, 26, 27]. However, these studies are based on agents with fixed morphs, without considering whether the body morphology of the agents is truly suitable for the tasks.

Thus, we propose **CompetEvo** that coevolves agent morphology and fighting tactics to integrate embodied morphological evolution into competitive games. We implement CompetEvo in two physical contact adversarial environments: `run-to-goal` [5] and `sumo` [2], shown in Fig. 5.14.

In contrast to the previous environments, our morphological arenas permit each agent to evolve their morphs before the commencement of each game. Therefore, the morphological evolution strategy is contemporaneously and continuously updated in synchronization with confrontational events. The goal of players in `run-to-goal` is to reach the red line behind the opponent, as illustrated in Fig. 5.14a. Both agents will try to prevent any passing; meanwhile, they will also try to break through the adversary's defense and reach the red line. The first to reach the red line wins the game. Moving toward the goal can get dense rewards in every step. At the end of the game, we set a large sparse winning reward (+1000) for the winner and punishment (−1000) for the loser in `run-to-goal`. In `sumo` shown in Fig. 5.14b,

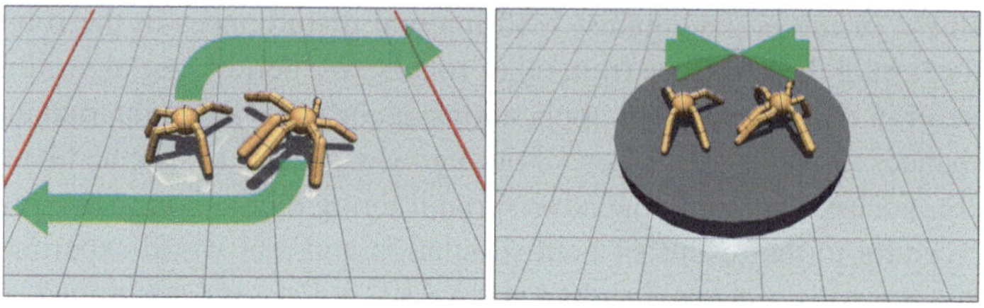

Fig. 5.14 Confrontation environments. `run-to-goal`: Agents are reset face-to-face and try to reach red lines behind the opponent as quickly as possible. `sumo`: Agents fight on dohyo and try to push the opponent out of the dohyo or knock opponent flat. (Reproduced with permission from [21])

players try to push the opponent out of the arena or a knockout to win the game. Quickly pushing the opponent outside the arena, staying at the center of the arena, and moving toward the opponent are possible ways to gain dense rewards. We set winning reward to +2000 and losing reward to −2000. Also, we penalize both sides (−1000) when a draw occurs to encourage confrontation; otherwise, agents might not learn aggressive behaviors.

5.4.1 Problem Formulation

We first employ a two-player Markov game [5] to formalize the problem, defined by: a set of states $\mathcal{S}$ describing the state of the world and possible states of both players, a set of actions of each player $\mathcal{A}_\alpha$ and $\mathcal{A}_\beta$ where we distinguish the agent and opponent by subscripts α and β, a joint transition function $\mathcal{T}: \mathcal{S} \times \mathcal{A}_\alpha \times \mathcal{A}_\beta \to \mathcal{S}$ determining distribution over the next states, reward functions $R_i : \mathcal{S} \times \mathcal{A}_\alpha \times \mathcal{A}_\beta \to \mathbb{R}, i \in \{\alpha, \beta\}$, contingent upon the current states and actions generated by each agent.

Previous work has studied two-player Markov games [5] and developed some interesting fighting skills. Based on this, we introduce morphological evolution, expanding the problem from the sole optimization of adversarial strategies to the joint optimization of confrontational morphology and fighting tactics, aiming to acquire improved morphologies tailored specifically for adversarial games. This promotion is highly challenging because it is not only evident in the morphology and skeletal structure of the agents but also manifested in their adept utilization of the evolved body to reinforce their combat skills, giving rise to novel fighting tactics.

To introduce morphological evolution, we formulate M_i to delineate the morph player $i \in \{\alpha, \beta\}$, which describes agent configurations like bone length and size, as well as joint limitations, and is involved in state sets $\mathcal{S}$. Here we define the coevolution combined policy as π that includes morph sub-policy with parameter θ and fighting tactics sub-policy with parameter ϕ. Given the predefined adversarial task and its corresponding rules, each player $i \in \{\alpha, \beta\}$ refines its policy $\pi_i(\theta; \phi)$, which generates a physical agent morphology and provides tactics for fighting. At the beginning of every game, evolvable players prepare an evolved morph M_i generated by morph sub-policy for combat. Each player aims to maximize its total expected return over time horizon T under confrontation. The cost function for agent i under morph M_i can be described as $J(\pi_i, M_i) = \mathbb{E}_{\pi_i, M_i}\left[\sum_{t=0}^{T} \gamma^t R_i\right]$, where γ is the discount factor for rewards, and T denotes the time horizon. A special case is when neither agent changes their morphs, and this problem degrades to a classical two-player game studied in [5, 29, 49].

5.4.2 Continuous Self-Practice Interactive Training

In the training of two-player games, a crucial aspect is to avoid significant imbalances in confrontations, as they may lead to divergent training strategies [5]: The disadvantaged agent struggles to acquire useful information in the competition, while the opponent, although in an advantageous position, lacks robustness in its strategies. Choosing appropriate opponents can address this issue. Our objective is to ensure that the agents' capabilities are continuously and collaboratively enhanced during confrontations. Therefore, the choice of an opponent becomes a problem that affects the stability of training, especially since the morphs of agents are also changeable in our environment.

To overcome this and refer to other two-player competition tasks, our training utilizes $\delta - Uniform$ opponent sampling, which ensures continual learning by training a policy that could consistently beat random previous versions of the opponent. Thus, the agent of the current policy from $\mathcal{P}_\alpha$ separately plays with multiple opponents of the previous policies from set $\mathcal{P}_\beta$. Let $\delta \in [0, 1]$ represent the percentage threshold applied to the oldest policy, determining its eligibility for potential sampling from the opponent player pool. At the commencement of each episode, we always choose the agent of the current policy and uniformly sample $\mathcal{N}$ historical opponent policies, depending on the specified δ threshold.

5.4.3 Morph Evolution and Tactics

We introduce the coevolution strategy into confrontation. First, we demonstrate how we encode agent morphology into parameters. Subsequently, we provide details of morphology optimization in adversarial games.

5.4.3.1 Morphology Encoding

We mainly use three types of species and their evolved derivatives. `ant`, `bug`, and `spider`, as shown in Fig. 5.15; then, based on these original morphs, we develop

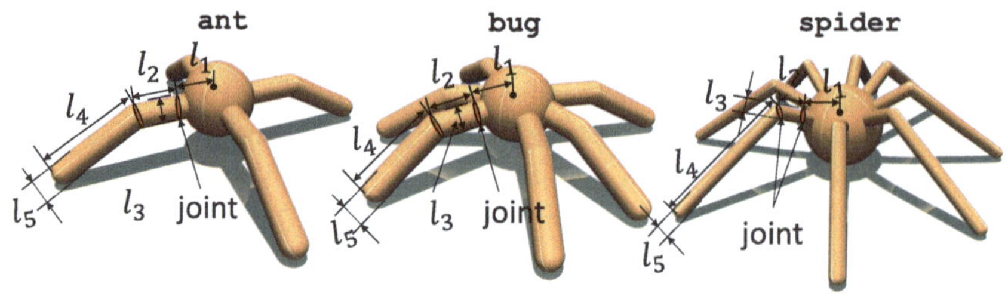

Fig. 5.15 Morphology encodings of three different agents: `ant`, `bug`, and `spider`. (Reproduced with permission from [21])

their evolvable versions: `evo-ant`, `evo-bug`, and `evo-spider`. The evolvable versions can adjust the morphological parameters of legs and the action capabilities corresponding to their morphs.

The agent morphology with fixed topology can be naturally represented using predefined variables [15]. We encode the designs of three animals using structural rules, illustrated in Fig. 5.15. All legs of animals share a similar structure: two joints and two limbs. The difference lies in that each species has a different number of legs: four for `ant`, six for `bug`, and eight for `spider`. Therefore, we use a vector with five parameters to describe one leg. In detail, l_2, l_3, l_4, l_5 denote the length and size of the thigh and lower leg, respectively, and l_1 indicates the distance from thigh to torso. To build links between morph and capability, the limb size maintains a linear relationship with the joint ability (moment arms, velocity, and force). In brief, the more robust limb can generate greater power. Subsequently, we encode agent design M into a vector $m = (l_1, ..., l_n)$. It is worth mentioning that variables in m are the scaling factor relative to the original agent but not the actual physical length. This is similar to a normalization of the real parameters, which makes morph optimization easier. Moreover, we limit the scaling of limbs to avoid a great disparity in size because our arenas have limited size and excessively large bodies may exceed the spatial constraints of the environment.

5.4.3.2 Morph and Tactics Joint Optimization in Confrontation

Thanks to the advancements in previous evolutionary algorithms [15, 44, 49], we now have a strong foundation for co-optimization. Unlike population-based bi-level optimization methods [40, 44], we take a more direct approach inspired by Transform2Act [49]. Morph design parameters are generated from the original settings x through the morph sub-policy network at the beginning of each episode and gradually converge with the growth of training generations (epochs).

We demonstrate self-practice coevolution optimization in Fig. 5.16 and Algorithm 5.2. We divide the process of adversarial coevolution into two stages: morph generation and arena confrontation. First, we illustrate how to introduce morphological factors into an adversarial game in Fig. 5.16. At the beginning of

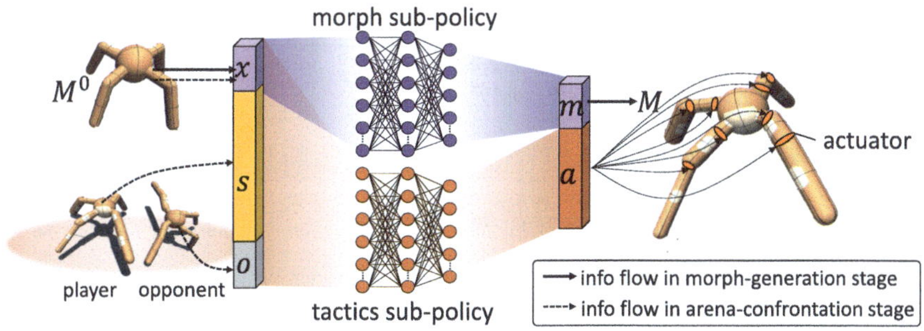

Fig. 5.16 Information flow in morph and tactics coevolution training. (Reproduced with permission from [21])

each game, we select policies for players and execute morphology generation to obtain morph pairs M_α and M_β to create an arena. The morph-generation stage is the first step of an episode, followed by the arena-confrontation stage.

More specifically, evolvable agents' morphology parameters m are generated by policy $\pi(\theta)$ and then create a morph M in the simulator. Note that for agents without morph evolution ability, we use fixed morphs and train only fighting tactics. Then combats start, and players concatenate ontological perception s, observation of opponent o, as well as morphology encoding m, generating actions through tactics sub-policy network. At the end of epochs, the latest policies of both players are collected and saved for later training. We sample data and update policies for two players by PPO, respectively, in every epoch.

5.4.4 Training Curriculum

5.4.4.1 Warming-up

Training from scratch is extremely challenging, especially when the agent lacks any fundamental skills. To expedite the training process and progress toward learning confrontational skills, it is essential to initially train the agent in basic behaviors such as walking. This facilitates both agents to move rapidly toward each other, generating physical contact and confrontational interaction. Our designed warm-up rewards give the agent a better tendency to learn basic skills, including less energy consumption, higher speed, and correct moving directions. We generally train about 100 epochs to guide agents in basic skills.

Furthermore, appropriate guidance during the warm-up phase can help avoid some strange phenomena during competition training. For instance, training without warming up could lead to `spider` learning the behavior of jumping over an opponent to avoid physical confrontation in `run-to-goal`. These behaviors are undesirable since they are beyond the ability dimension of common players and

Algorithm 5.2 Confrontation algorithm for coevolving agents

Input: initial policies π_α^0 and π_β^0, original morph M_α^0, and M_β^0, opponent sampling factor δ
Parameter: memory $\mathcal{M}$, policy pools $\mathcal{P}_\alpha$ and $\mathcal{P}_\beta$
Output: policies π_α and π_β

$\mathcal{P}_\alpha \leftarrow \emptyset, \mathcal{P}_\beta \leftarrow \emptyset$
$\mathcal{P}_\alpha \leftarrow \mathcal{P}_\alpha \bigcup \{\pi_\alpha^0\}, \mathcal{P}_\beta \leftarrow \mathcal{P}_\beta \bigcup \{\pi_\beta^0\}$
while *not reaching maximum generation* **do**
  for player i in $\{\alpha, \beta\}$ **do**
    $\mathcal{M} \leftarrow \emptyset$
    Define opponent player $j \neq i$ and $j \in \{\alpha, \beta\}$
    Sample policies π_i, π_j from $\mathcal{P}_i, \mathcal{P}_j$
    while *episode not over* **do**
      if *at morph-generation stage* **then**
        Get M_i, M_j by $m_i \sim \pi_i(\theta), m_j \sim \pi_j(\theta)$
        Create morph arena by M_i, M_j
        $r_i \leftarrow 0$; store (r_i, m_i) into $\mathcal{M}$
      else if *at arena-confrontation stage* **then**
        $a_i \sim \pi_i(\phi), a_j \sim \pi_j(\phi)$
        $s' \leftarrow \mathcal{T}(s, a_i, a_j); s \leftarrow s'$
        $r_i \leftarrow R_i(s, a_i, a_j)$
        Store (r_i, a_i, m_i, s) into $\mathcal{M}$
      end if
    end while
    Update π_i with PPO using data in $\mathcal{M}$
    $\mathcal{P}_i \leftarrow \mathcal{P}_i \bigcup \{\pi_i\}$
  end for
end while
return π_α, π_β

are prone to premature strategy divergence. To avoid this, we guide agents to move forward as close to the ground as possible to prepare for the later physical confrontation.

5.4.4.2 Rewards Annealing

When the agents can move toward each other and make physical contact, confrontation training begins in earnest. At the end of every episode, we give winners a huge positive fighting reward and a negative one for losers. Nonetheless, fighting rewards in two-player games are too sparse for reinforcement learning, which makes training extremely challenging because the agent has to learn the correct actions with limited positive feedback. Training with only sparse rewards often leads to issues like signal delay, suboptimal problem [36], and high variance.

Therefore, we use both dense rewards and sparse rewards simultaneously to overcome exploration difficulty. The former encourages basic skills learning, and the latter provides stimulation for confrontation. For dense rewards design, we extend the reward settings from the warming-up phase to enhance the fundamental skills.

To balance dense rewards R_d and sparse rewards R_s, we use an annealing factor κ to make a trade-off. κ varies depending on the current generation t and predefined termination generation T_t:

$$R = \kappa R_d + (1 - \kappa) R_s, \ \kappa = \max\left(\frac{T_t - t}{T_t}, 0\right). \tag{5.17}$$

In the beginning, dense rewards dominate the training direction and then gradually reduce the influence until κ declines to zero. After the termination generation T_t, only sparse rewards work. We generally set termination generation as the number of maximum training iterations, or half of it.

5.4.5 Experiments

To validate the performance of agents with distinct morphs, we set cross-antagonism for agents with the same training iterations and use their win rates over one hundred rounds to represent their abilities.

5.4.5.1 Coevolution in Self-Practice

Evolvable agents simultaneously optimize their designs and fighting tactics through self-practice. During the training process, the design parameters' varying tendencies and the final morphs after convergence are illustrated in Fig. 5.17.

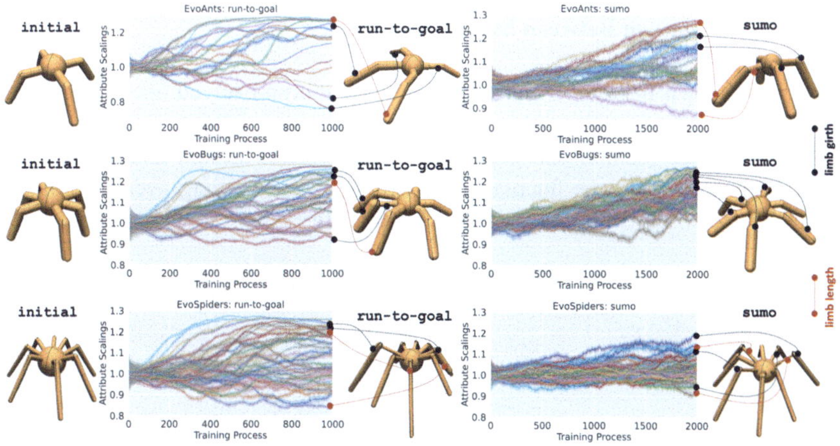

Fig. 5.17 Morph parameter changes of evolvable agents in the training process and their final evolved morphologies. (Reproduced with permission from [21])

The trend can be inferred by observing the variations. In `run-to-goal` competition, agents tend to evolve much more robust limbs toward the moving direction. This tendency helps to improve the stability of the heading and thus resist the interference of the opponent, as shown in the middle column in Fig. 5.17. On the other hand, thighs tend to be thinner due to energy efficiency. This is because the lower legs play a more important role in propelling the agent than the thighs, and thighs degenerate due to energy saving. The right column in Fig. 5.17 illustrates that parameters growing in `sumo` are more concentrated and show the tendency to increase. Agents tend to develop brawny lower legs, and only very few limbs become thinner. This feature can be found on most legs because agents with stronger limbs can provide much more powerful impulsive force during confrontation to cope with threats from all directions.

5.4.5.2 Effectiveness of Coevolution

As aforementioned, there are three species in our experiments: `ant`, `bug`, and `spider`, corresponding to their three original morphs and the three evolved morphs derived from them. To validate that morphological evolution can improve the capabilities of fighting, we first conduct experiments and set confrontations on symmetric species, between the original agents and their evolved versions. The expectation is that agents with evolved morphologies will achieve a higher win rate compared to their original design. Moreover, to further validate the effectiveness and generalization performance of evolved agents, we also conduct adversarial experiments among asymmetric species. We pit the original agent and the evolved version of one species against other species, comparing their respective winning rates.

5.4.5.3 Comparison Between Symmetric Species

Results are shown in Fig. 5.19. The win rates of morph-evolved agents are much higher than those of agents with original morphologies in all six scenarios, which indicates that our method can generate a more suitable design and strategy for confrontation. Besides, the fluctuations of the win rates varying over the training epochs are also shown in Fig. 5.18. The blue lines represent the win rate of

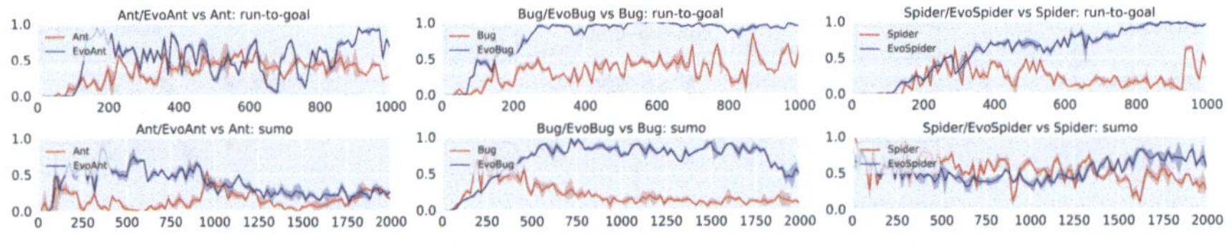

Fig. 5.18 Win rates of evolvable and original agents in competition between symmetric species. (Reproduced with permission from [21])

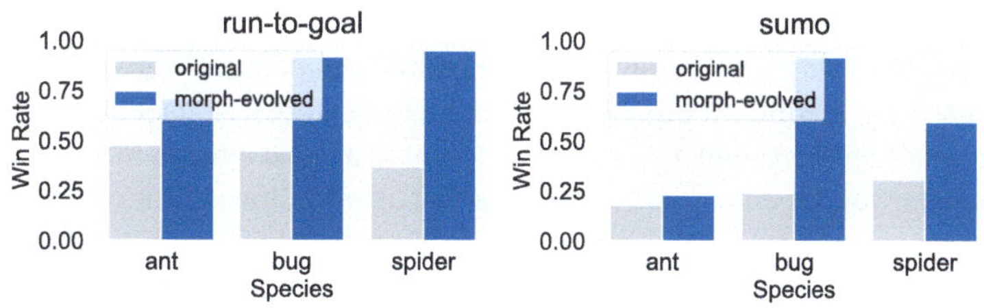

Fig. 5.19 Win rate comparison between evolvable agents and original agents when facing a different species. (Reproduced with permission from [21])

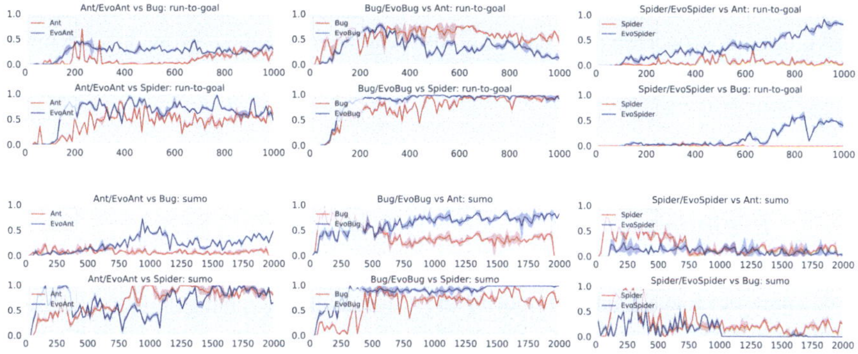

Fig. 5.20 Win rates of evolvable and original agents in competition between asymmetric species. The vertical axis denotes the win rate, and the horizontal axis denotes the iteration of the training process. (Reproduced with permission from [21])

evolved agents fighting against original agents, while the red lines represent the win rates between two original agents. Nearly all evolvable agents consistently maintain a dominant position throughout the majority of the training process over two adversarial scenarios, which demonstrates the robustness and applicability of our method (Fig. 5.19).

5.4.5.4 Comparison Between Asymmetric Species

Further evidence can be found in asymmetric species. We select evolvable and original agents of the same species, engaging them in battles against agents from other species, and compare the resulting changes in their respective winning rates. According to results shown in Fig. 5.20, evolvable agents maintain higher win rates than those of their original morph in the majority of scenarios. Besides, in nearly half of the scenarios, the winning rates of evolvable agents get a promotion and eventually surpass those of their original morphs. There are also some scenarios where the effects are not pronounced. For example, in experiments of `spider` and evo-spider in sumo task, both of them get low win rates. This is because the spider agents have too many legs, requiring more precise cooperation between joints for movement, often leading to instability when facing external impacts. Additionally, the thin legs of spider agents inherently put them at a disadvantage in limb contact.

The main evidence consistently indicates that agents allowing for morphological evolution possess stronger capabilities. Nonetheless, this does not mean that evolved agents can defeat all physically weaker opponents of other species. There is only one failure case: evo-bug versus ant in run-to-goal. ant is much smaller, which helps it evade a frontal attack from a formidable opponent. It can hide under the evo-bug, lifting the opponent off the ground and carrying it to the goal.

Although competitions between different species are inherently unfair, our results reveal that in competitions between asymmetric species, employing a coevolution approach enables naturally disadvantaged agents to grow up with more robust morphologies, thereby enhancing their winning rates.

5.4.6 Typical Examples Illustrating the Roles of Morphological Evolution

In this part, our focus is on highlighting the remarkable emergent behaviors resulting from morphological evolution. We conduct competitions by cross-confrontations among all six types of agents.

The morphological evolution leads to interesting behaviors in the combat process. For example, evolvable agents tend to develop more robust limbs as they practice themselves, which gives them an unequivocal advantage over their original design. Here, we present four observed emergent behaviors resulting from the evolution of morphology: throwing, wrestling, standing, and defending, shown in Fig. 5.21a. In run-to-goal, we observe the behavior of throwing: Agents evolve larger front legs, utilizing the inertia of the front legs to attempt to throw themselves forward, which is much faster than the original morphs. This phenomenon can be observed in all species of evolvable agents shown in Fig. 5.17.

Besides, agents with morphological evolution exhibit limbs with greater strength, sufficient to employ techniques resembling wrestling moves to overturn opponents on the ground, as shown in Fig. 5.21b. The evo-bug develops three very strong legs, clamping onto one of the opponent's legs and then directly overturning them to the ground.

In task sumo, we also observed interesting behaviors corresponding to tactical strategies. In evo-spider versus evo-spider, spiders' numerous legs make it challenging to maintain coordination and balance to stand. Through morphological evolution, the spider evolved two to three relatively long legs among its many legs, forming a triangular support with the shorter legs. This arrangement prevents them from losing balance in most situations, shown in Fig. 5.21c.

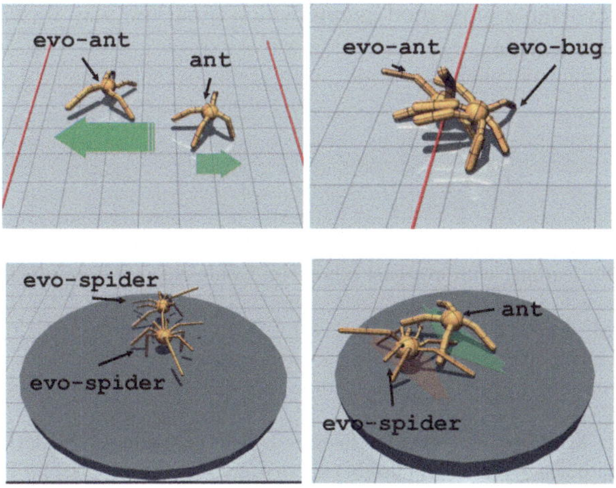

Fig. 5.21 Emerging behaviors related to morphological evolution: (left-up): Throwing out itself with robust limbs for higher speeds; (right-up): Wrestling strategy to disrupt the rival's balance; (left-down): Standing aided by limbs of varying lengths; (right-down): Defending by expanding the limbs support angle. (Reproduced with permission from [21])

Furthermore, morphological evolution gives rise to richer defensive tactics, illustrated in Fig. 5.21d. `evo-spider` evolves various lengths of legs. Longer side legs, as well as shorter front and back legs, are more advantageous for adopting a defensive posture. When facing opponents, `evo-spider` spread their legs to create a more stable point of force, resisting opponents' impacts, which can facilitate their defense tactics.

5.5 Summary

This section has explored keys of competition in embodied multi-agent system. Learning from competitive information first demonstrates how leveraging interactions among agents with the same goals that can stimulate potential from competition. Then, adversarial skill learning is described in detail, answering how to use contrast learning for improving resilience in dynamic environments with external competition, where adversarial training enables agents to develop manipulation skills that withstand perturbations and unexpected challenges. Eventually, the coevolution of policy and body in competitive two-player games is introduced that reveals the mutual shaping of physical and morphological embodiment and cognitive capabilities, emphasizing how competition drives morphologically adapted agents for specific tasks.

In summary, this section provides several different aspects of competition in embodied multi-agent system. Our examples include simulated and physical robots, especially the evolvable embodied agents which point the way forward for multi-agent competitive learning.

References

1. Ahn M, Zhu H, Hartikainen K, Ponte H, Gupta A, Levine S, Kumar V (2020) Robel: Robotics benchmarks for learning with low-cost robots. In: Conference on Robot Learning, pp 1300–1313. PMLR
2. Al-Shedivat M, Bansal T, Burda Y, Sutskever I, Mordatch I, Abbeel P (2018) Continuous adaptation via meta-learning in nonstationary and competitive environments. In: 6th International Conference on Learning Representations, ICLR 2018 - Conference Track Proceedings (March), pp 1–21
3. Andrychowicz M, Wolski F, Ray A, Schneider J, Fong R, Welinder P, McGrew B, Tobin J, Abbeel OP, Zaremba W (2017) Hindsight experience replay. Adv Neural Inf Proc Syst, 5048–5058
4. Asada M, Noda S, Tawaratsumida S, Hosoda K (1996) Purposive behavior acquisition for a real robot by vision-based reinforcement learning. Mach Learn 23(2–3):279–303
5. Bansal T, Pachocki J, Sidor S, Sutskever I, Mordatch I (2018) Emergent complexity via multi-agent competition. In: 6th International Conference on Learning Representations, ICLR 2018 - Conference Track Proceedings
6. Bowman M, Li S, Zhang X (2019) Intent-uncertainty-aware grasp planning for robust robot assistance in telemanipulation. In: 2019 International Conference on Robotics and Automation (ICRA), pp 409–415. IEEE
7. Brockman G, Cheung V, Pettersson L, Schneider J, Schulman J, Tang J, Zaremba W (2016) OpenAI Gym
8. Chen T, Kornblith S, Norouzi M, Hinton G (2020) A simple framework for contrastive learning of visual representations. In: 37th International Conference on Machine Learning, ICML 2020 PartF16814, 1575–1585
9. Creswell A, White T, Dumoulin V, Arulkumaran K, Sengupta B, Bharath AA (2018) Generative adversarial networks: An overview. IEEE Signal Process Mag 35(1):53–65. https://doi.org/10.1109/MSP.2017.2765202
10. de Witt CS, Gupta T, Makoviichuk D, Makoviychuk V, Torr PHS, Sun, M., Whiteson S (2020) Is Independent Learning All You Need in the StarCraft Multi-Agent Challenge? http://arxiv.org/abs/2011.09533
11. Gao T, Yao X, Chen D (2021) SimCSE: Simple Contrastive Learning of Sentence Embeddings. In: EMNLP 2021 - 2021 Conference on Empirical Methods in Natural Language Processing, Proceedings pp 6894–6910. https://doi.org/10.18653/v1/2021.emnlp-main.552
12. Ghosh S, Berkenkamp F, Ranade G, Qadeer S, Kapoor A (2018) Verifying controllers against adversarial examples with bayesian optimization. In: 2018 IEEE International Conference on Robotics and Automation (ICRA), pp 7306–7313. IEEE
13. Gu Z, Jia Z, Choset H (2019) Adversary a3c for robust reinforcement learning. arXiv preprint arXiv:1912.00330
14. Gullapalli V, Franklin JA, Benbrahim H (1994) Acquiring robot skills via reinforcement learning. IEEE Control Syst Mag 14(1):13–24
15. Ha D (2019) Reinforcement learning for improving agent design. Artif Life 25(4):352–365. https://doi.org/10.1162/artl_a_00301
16. Haarnoja T, Moran B, Lever G, Huang SH, Tirumala D, Wulfmeier M, Humplik J, Tunyasuvunakool S, Siegel NY, Hafner R, Bloesch M, Hartikainen K, Byravan A, Hasenclever L, Tassa Y, Sadeghi F, Batchelor N, Casarini F, Saliceti S, Game C, Sreendra N, Patel K, Gwira M, Huber A, Hurley N, Nori F, Hadsell R, Heess N (2023) Learning Agile Soccer Skills for a Bipedal

Robot with Deep Reinforcement Learning. arXiv preprint arXiv:2304.13653. http://arxiv.org/abs/2304.13653
17. Haarnoja T, Zhou A, Abbeel P, Levine S (2018) Soft actor-critic: Off-policy maximum entropy deep reinforcement learning with a stochastic actor. In: International Conference on Machine Learning, pp 1861–1870
18. Hu J, Hu S, Liao Sw (2021) Policy Regularization via Noisy Advantage Values for Cooperative Multi-agent Actor-Critic methods. http://arxiv.org/abs/2106.14334
19. Huang K, Chen J, Oyekan J (2021) Decentralised aerial swarm for adaptive and energy efficient transport of unknown loads. Swarm Evol Comput 67:100957. https://doi.org/10.1016/j.swevo.2021.100957
20. Huang K, Chen J, Oyekan J, Zhang X (2022) Bio-inspired Multi-agent Model and Optimization Strategy for Collaborative Aerial Transport. In: Lecture Notes in Electrical Engineering 801 LNEE, pp 591–598. https://doi.org/10.1007/978-981-16-6372-7_64
21. Huang K, Guo D, Zhang X, Ji X, Liu H (2024) CompetEvo: Towards Morphological Evolution from Competition. In: K. Larson (ed.) Proceedings of the Thirty-Third International Joint Conference on Artificial Intelligence, IJCAI-24, pp 85–93. International Joint Conferences on Artificial Intelligence Organization. https://doi.org/10.24963/ijcai.2024/10
22. Huang K, Guo D, Zhang X, Ji X, Liu H (2024) Stimulate the potential of robots via competition. arXiv preprint arXiv:2403.10487
23. Jian P, Yang C, Di Guo, H.L., Sun F (2021) Adversarial skill learning for robust manipulation. In: Proceedings - IEEE International Conference on Robotics and Automation 2021-May, pp 2555–2561. https://doi.org/10.1109/ICRA48506.2021.9561379
24. Lillicrap TP, Hunt JJ, Pritzel A, Heess N, Erez T, Tassa Y, Silver D, Wierstra D (2016) Continuous control with deep reinforcement learning. In: ICLR (Poster)
25. Littman ML (1994) Markov games as a framework for multi-agent reinforcement learning. In: Machine Learning Proceedings 1994, pp 157–163. Elsevier
26. Liu S, Lever G, Merel J, Tunyasuvunakool S, Heess N, Graepel T (2019) Emergent coordination through competition. In: 7th International Conference on Learning Representations, ICLR 2019
27. Liu S, Lever G, Wang Z, Merel J, Eslami SM, Hennes D, Czarnecki WM, Tassa Y, Omidshafiei S, Abdolmaleki A, Siegel NY, Hasenclever L, Marris L, Tunyasuvunakool S, Song HF, Wulfmeier M, Muller P, Haarnoja T, Tracey B, Tuyls K, Graepel T, Heess N (2022) From motor control to team play in simulated humanoid football. Sci Robot 7(69). https://doi.org/10.1126/scirobotics.abo0235
28. Luo G, Zhang H, He H, Li J, Wang FY (2021) Multiagent adversarial collaborative learning via mean-field theory. IEEE Trans Cybern 51(10):4994–5007. https://doi.org/10.1109/TCYB.2020.3025491
29. Maruan Al-S, Trapit B, Yura B, Ilya S, Igor M, Pieter A (2013) Continuous adaptation via meta-learning in nonstationary and competitive environments (March), 1–21
30. Morimoto J, Doya K (2005) Robust reinforcement learning. Neural Comput 17(2):335–359
31. Perolat J, Scherrer B, Piot B, Pietquin O (2015) Approximate dynamic programming for two-player zero-sum Markov games. In: International Conference on Machine Learning, pp 1321–1329
32. Petersen IR, Ugrinovskii VA, Savkin AV (2012) Robust control design using h methods. Springer Science & Business Media, Berlin
33. Pinto L, Davidson J, Gupta A (2017) Supervision via competition: Robot adversaries for learning tasks. In: 2017 IEEE International Conference on Robotics and Automation (ICRA), pp 1601–1608. IEEE
34. Pinto L, Davidson J, Sukthankar R, Gupta A (2017) Robust adversarial reinforcement learning. In: International Conference on Machine Learning, pp 2817–2826
35. Porav H, Maddern W, Newman P (2018) Adversarial training for adverse conditions: Robust metric localisation using appearance transfer. In: 2018 IEEE International Conference on Robotics and Automation (ICRA), pp 1011–1018. IEEE

36. Riedmiller M, Hafner R, Lampe T, Neunert M, Degrave J, Van De Wiele T, Mnih V, Heess N, Springenberg T (2018) Learning by playing - Solving sparse reward tasks from scratch. In: 35th International Conference on Machine Learning, ICML 2018, vol 10, pp 6910–6919
37. Schulman J, Moritz P, Levine S, Jordan MI, Abbeel P (2016) High-dimensional continuous control using generalized advantage estimation. In: 4th International Conference on Learning Representations, ICLR 2016 - Conference Track Proceedings
38. Silver D, Huang A, Maddison CJ, Guez A, Sifre L, Van Den Driessche G, Schrittwieser J, Antonoglou I, Panneershelvam V, Lanctot M, Dieleman S, Grewe D, Nham J, Kalchbrenner N, Sutskever I, Lillicrap T, Leach M, Kavukcuoglu K, Graepel T, Hassabis D (2016) Mastering the game of Go with deep neural networks and tree search. Nature 529(7587):484–489. https://doi.org/10.1038/nature16961
39. Silver D, Hubert T, Schrittwieser J, Antonoglou I, Lai M, Guez A, Lanctot M, Sifre L, Kumaran D, Graepel T, Lillicrap T, Simonyan K, Hassabis D (2018) A general reinforcement learning algorithm that masters chess, shogi, and Go through self-play. Science 362(6419):1140–1144. https://doi.org/10.1126/science.aar6404
40. Sims K (1994) Evolving virtual creatures. In: Proceedings of the 21st Annual Conference on Computer Graphics and Interactive Techniques, SIGGRAPH 1994, pp 15–22. https://doi.org/10.1145/192161.192167
41. Slotine JJE (1985) The robust control of robot manipulators. Int J Robot Res 4(2):49–64
42. Tessler C, Efroni Y, Mannor S (2019) Action robust reinforcement learning and applications in continuous control. In: International Conference on Machine Learning, pp 6215–6224
43. Wang J, Yu L, Zhang W, Gong Y, Xu Y, Wang B, Zhang P, Zhang D (2017) IRGAN: A minimax game for unifying generative and discriminative information retrieval models. In: SIGIR 2017 - Proceedings of the 40th International ACM SIGIR Conference on Research and Development in Information Retrieval pp 515–524. https://doi.org/10.1145/3077136.3080786
44. Wang T, Zhou Y, Fidler S, Ba J (2019) Neural graph evolution: Towards efficient automatic robot design. In: 7th International Conference on Learning Representations, ICLR 2019
45. Wulfmeier M, Bewley A, Posner I (2018) Incremental adversarial domain adaptation for continually changing environments. In: 2018 IEEE International Conference on Robotics and Automation (ICRA), pp 1–9. IEEE
46. Yang Y, Luo R, Li M, Zhou M, Zhang W, Wang J (2018) Mean field multi-agent reinforcement learning. In: 35th International Conference on Machine Learning, ICML 2018 12, 8869–8886
47. Yu C, Velu A, Vinitsky E, Gao J, Wang Y, Bayen A, Wu Y (2021) The Surprising Effectiveness of PPO in Cooperative, Multi-Agent Games. http://arxiv.org/abs/2103.01955
48. Yu C, Velu A, Vinitsky E, Gao J, Wang Y, Bayen A, Wu Y (2022) The surprising effectiveness of PPO in cooperative multi-agent games. Adv Neural Inf Proc Syst 35. https://doi.org/10.48550/arXiv.2103.01955
49. Yuan Y, Song Y, Luo Z, Sun W, Kitani KM (2022) Transform2Act: learning a transform-and-control policy for efficient agent design. In: ICLR 2022 - 10th International Conference on Learning Representations

Chapter 6
Large Language Model for Embodied Multi-Agent System

Abstract In this chapter, we extend the framework of embodied multi-agent system from the learning-based method to the training-free strategy based on the large language model (LLM). We consider a more challenging heterogeneous ad hoc teamwork collaboration problem where an ad hoc robot joins an existing heterogeneous team for a shared goal. Specifically, the ad hoc robot collaborates with unknown teammates without prior coordination, and it is expected to generate an appropriate cooperation policy to improve the efficiency of the whole team. We first discuss and analyze the problem of ad hoc teamwork collaboration problem. Then we introduce the proposed extended framework based on LLM. We leverage the remarkable potential of the large language model (LLM) to establish a decentralized heterogeneous ad hoc teamwork collaboration framework that focuses on generating reasonable policy for an ad hoc robot to collaborate with original heterogeneous teammates. A training-free hierarchical dynamic planner is developed using the LLM together with the newly proposed Interactive Reflection of Thoughts (IRoT) method for the ad hoc agent to adapt to different teams. We also build a benchmark testing dataset to evaluate the proposed framework in the heterogeneous ad hoc multi-agent tidying-up task. Extensive comparison and ablation experiments are conducted in the benchmark to demonstrate the effectiveness of the proposed framework.

6.1 Introduction

Imagine after a natural disaster such as an earthquake or hurricane, a team of robots is dispatched for the rescue task. Since the situation of a disaster site is complex, robots of different capabilities may be required for the rescue. These robots are likely to be brought from different places and thus arrive at the site at different times. The coming robot does not have any prior information on existing teammates, and it is expected to collaborate efficiently and robustly with previously unknown teammates for the same goal. This scenario describes a typical heterogeneous ad hoc teamwork, and the new coming robot is called an ad hoc robot. Specifically, a heterogeneous ad hoc teamwork collaboration is demonstrated in Fig. 6.1, where

H. Liu et al., *Embodied Multi-Agent Systems*, Machine Learning: Foundations, Methodologies, and Applications, https://doi.org/10.1007/978-981-96-5871-8_6

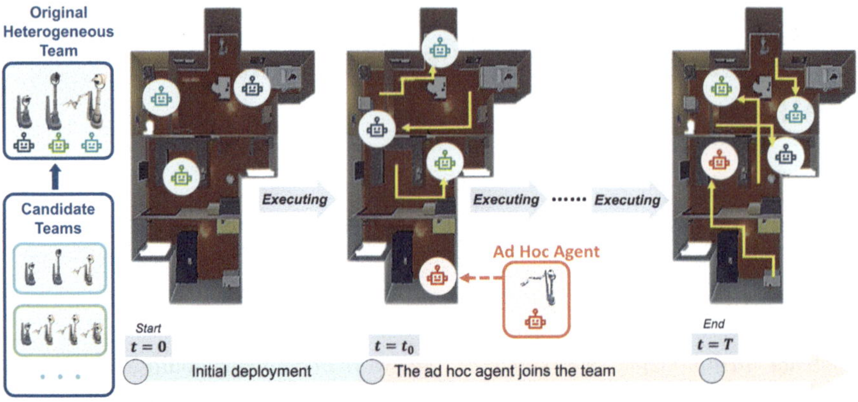

Fig. 6.1 The illustration of heterogeneous ad hoc teamwork collaboration

heterogeneous robots of different capabilities can compose any team, and the original heterogeneous team collaborates to execute a task. An ad hoc robot could join this team at any time from any location, and then a heterogeneous ad hoc team could collaborate to achieve the goal.

During the past years, the multi-robot collaboration task has been widely investigated, and a bunch of multi-agent embodied tasks are proposed where multiple agents learn proper strategies to collaborate efficiently [11, 12, 14, 15, 29–31, 36] and solve complex embodied tasks [17, 23]. All these works only consider homogeneous agents with the same capabilities. However, in real-world applications, the robots may be faced with more complicated situations such as seismic sensing and planetary exploration. It is necessary to leverage robots with different capabilities to accomplish the task better [8, 13, 24, 25, 28]. Meanwhile, the ad hoc teamwork collaboration is an important problem in the heterogeneous multi-robot collaboration, which has been rarely addressed. In the heterogeneous ad hoc team collaboration task, an ad hoc robot has to collaborate with a group of unknown teammates without any prior coordination. An appropriate cooperation policy should be generated for the ad hoc robot to improve the efficiency of the whole team and avoid interfering with the effective collaboration among original teammates, which makes it an extremely challenging task.

The formal definition of ad hoc teamwork is proposed in [27] and has attracted a surge of studies in the past years [1–4, 7, 37]. However, most of these works discuss ad hoc teamwork collaboration as a theoretical problem under relatively ideal conditions in some simple 2D environments. Reference [30] considers the ad hoc setting in the 3D embodied environment. Still, it only discusses the situation in which the number of agents in the training process differs from that in the testing process without thinking of an ad hoc agent adapting to changing teams. Furthermore, prior studies mostly focus on the situation in which the ad hoc agent joins the team at the beginning of the task, but in practical scenes, the ad hoc agent is likely to join the team at any time. Therefore, existing ad hoc teamwork studies are difficult to apply directly to practical robotic applications. In this chapter, we focus on a more general ad hoc teamwork setting in which the ad hoc agent can join any team, at any time, and from any location.

On the other hand, the Large Language Model (LLM) has recently shown promising performance in natural language understanding, reasoning, task planning, as well as decision-making in robotics [10, 22]. And there emerge a lot of works trying to leverage the LLM to perform high-level planning for robots [5, 9, 10, 16, 19, 20, 35]. To generate the reasonable policy for an ad hoc robot to collaborate with a group of unknown teammates without any prior coordination, we propose to leverage the LLM to perform robust high-level reasoning and planning for the ad hoc agent to effectively collaborate with any team at any time. A novel framework is established for heterogeneous ad hoc teamwork collaboration. This chapter discusses the following contents:

- **Framework:** We propose a decentralized heterogeneous ad hoc teamwork collaboration framework, which supports the ad hoc agent to join any previously unknown teammates at any time from any location.
- **Method:** A hierarchical dynamic planning model based on the LLM is developed, in which the Interactive Reflection of Thought (IRoT) method is proposed for the LLM to generate the sub-task and a subskill planner is developed to predict the next subskill for the ad hoc agent to execute. This model can adapt to teammates with different capabilities and policies.
- **Validation:** We build a benchmark testing dataset based on ProcTHOR-10K [6] to evaluate the proposed ad hoc teamwork collaboration framework in the heterogeneous multi-agent tidying-up task. Experiments are conducted in both the simulation and physical scenarios, illustrating the effectiveness of the proposed heterogeneous ad hoc collaboration framework.

6.2 Ad Hoc Teamwork Problem in Embodied Multi-Agent System

In this chapter, we consider a more challenging heterogeneous ad hoc teamwork collaboration problem. We examine a scenario involving an ad hoc agent collaborating with unfamiliar teammates at any time, from any location. In the absence of prior coordination, the agent must adapt to the original team by communicating basic information with teammates and observing teammates' behaviors.

Given a goal $\mathcal{G}$ and an existing team $\mathcal{T} = \{T_1, \cdots, T_N\}$, where T_n is the n-th embodied agent and N is the number of teammates[1], we denote the policy set as $\Pi = \{\pi_1, \pi_2, \cdots, \pi_N\}$, where π_n is the policy of the n-th embodied agent T_n.

Though the multi-agent collaboration is still not fully addressed, there exists a lot of work that provides many feasible solutions. Therefore, in this chapter, we

[1] In the task setting of this chapter, the teammates are all agents or robots and do not contain humans.

assume we have already obtained a feasible policy set $\bar{\Pi}$ for the existing team, and the objective is to design the policy for an ad hoc agent T_0, which may enter the environment at any timestep (denoted as $t_0 \geq 0$), to interact with existing agents in $\mathcal{T}$ to complete the goal $\mathcal{G}$. The policy of the ad hoc agent T_0 is denoted as π_0, of which optimal solution can be obtained as

$$\pi_0^* = \arg\max_{\pi_0} \quad U_{t \geq t_0}(\mathcal{G}, \mathcal{T} \cup T_0, \bar{\Pi} \cup \pi_0), \tag{6.1}$$

where $U_{t \geq t_0}()$ is the utility function which is dependent on the goal, agents, and policies. The subscript $t \geq t_0$ means it is accumulated from the time instant t_0. Note that in the above equation, $\bar{\Pi}$ is the fixed policy for existing teammates, and the utility function could be different from the one that is used to optimize $\bar{\Pi}$. This guarantees the existing team $\mathcal{T}$ is indeed a black box to the ad hoc agent T_0.

Note that even though we set $t_0 = 0$, the problem in Eq. 6.1 is totally different from the conventional multi-agent collaboration. Because in our ad hoc setting (Eq. 6.1), we can only optimize the policy of the ad hoc agent, while other agents keep their policies unchanged. In addition, the ad hoc agent has no prior information about the existing teammates $\mathcal{T}$. During the collaboration procedure, the ad hoc agent should observe the situation, communicate with its teammates, and try to adapt to their behaviors, which is a challenging zero-shot collaboration problem.

Further, we give more detailed descriptions of the embodied agent, which clarify the boundary of the problem in this work. In the following, we do not differentiate the ad hoc agent ($n = 0$) and the existing teammates ($n = 1, 2, \cdots, N$).

For the n-th embodied agent (for $n = 0, 1, 2, \cdots, N$), we define a 5-tuple $\mathcal{W}_n = \{\mathcal{S}_n, \mathcal{M}_n, C_n, \mathcal{A}_n, \mathcal{H}_n\}$, which is illustrated as:

1. $\mathcal{S}_n$: State space, which includes the observations, the status (location, orientation, power assumption, ...), etc.
2. $\mathcal{M}_n$: Messages space, which includes the communication messages with other available agents
3. C_n: Capability space, which includes the agent capability of action, payload, and resources
4. $\mathcal{A}_n$: Action space, which includes the practically execution actions
5. $\mathcal{H}_n$: Memory space, which includes historical data containing states, actions, messages, etc.

Note that the above five sets are mutually coupled. For example, the action $\mathcal{A}_n$ is obviously influenced by the capability space C_n. In addition, $\mathcal{W}_{n_1}$ and $\mathcal{W}_{n_2}$ can be significantly different for $n_1 \neq n_2$, which is a prerequisite for heterogeneous ad hoc collaboration. Based on the above setting, we formalize the corresponding policy as the mapping

$$\pi_n(\cdot) : \{\mathcal{S}_n, \mathcal{M}_n, C_n, \mathcal{H}_n\} \rightarrow \Delta(\mathcal{A}_n), \tag{6.2}$$

where Δ represents probability distribution over the corresponding set.

Before closing this section, we emphasize that the above problem setting strictly complies with the assumptions summarized in [18]: (1) No prior coordination. The ad hoc agent is expected to cooperate with its teammates when the task begins without any prior opportunities to establish or specify mechanisms for coordination. (2) No control over teammates: The ad hoc agent cannot change the properties of the environment, the teammates' policies, and communication protocols. It has to reason and act under the given conditions. The properties of the environment include the observability level, the modality of the observation information the agent obtains, the categories of objects that may exist in the environment, and so on, which is different from the state of the environment.

6.3 Ad Hoc Framework in Embodied Multi-Agent System

An overview of the proposed decentralized heterogeneous ad hoc teamwork collaboration framework is demonstrated in Fig. 6.2, which plans and generates the action decision only with the state information of the ad hoc agent and the communication messages from teammates. The decentralized framework means that there does not exist a central node to generate the action decision for all teammates and the ad hoc agent by collecting all their state information. The visual semantic perception module is responsible for generating the top-down semantic map and the hierarchical scene graph of the environment. An LLM-based hierarchical planner, which consists of a dynamic sub-task planner and dynamic subskill planner, is used for generating a coordination policy for the ad hoc agent. Specifically, the dynamic sub-task planner generates the next sub-task for the ad hoc agent to perform with a newly proposed Interactive Reflection of Thought (IRoT) mechanism. Then, the dynamic subskill planner generates the next subskill to execute based on the generated sub-task, and the subskill execution module generates the corresponding low-level actions for the ad hoc agent to execute the subskill. Finally, the state assessment module evaluates the action state after the execution and generates effective information to help replan the sub-task and subskill. The ad hoc agent transmits and receives the information with the original team through the communication module.

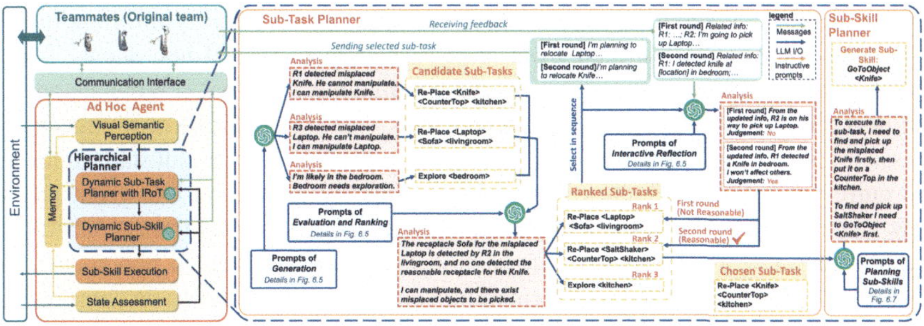

Fig. 6.2 The overview of the proposed ad hoc teamwork framework

We employ the proposed ad hoc framework in the tidying-up task setting, in which multiple agents need to find all misplaced objects and put them in proper receptacles. Specifically, receptacles refer to objects that other objects can be placed on their surface or inside. The misplaced object refers to the object that is put in the unreasonable receptacle according to common sense. In the tidying-up task, agents need to infer whether the detected object is misplaced and infer the reasonable receptacle to place it. Eventually, agents pick up all misplaced objects and put them in reasonable locations.

6.3.1 Preliminary Information

For convenience, we describe the needed information in the proposed framework under the tidying-up setting in advance as follows:

- State information ($\mathcal{S}_0$): the location of the agent, the object-grasping state, the observation, and key detection information, as well as the task-relevant state in the current timestep
- Message information ($\mathcal{M}_0$): all messages the agent has sent and received in the current timestep
- Capability information ($\mathcal{C}_0$): agent height, manipulation ability, payload (the maximum mass of objects that can be picked up), and the total number of steps the battery power can initially support
- Action information ($\mathcal{A}_0$): the action the agent takes in the current timestep
- Memory information ($\mathcal{H}_0$): all state, message, and action information in all previous timesteps.

6.3.2 Communication

The ad hoc agent utilizes the communication module to exchange effective information with the original team. When the ad hoc agent joins the original team, it will broadcast its state and capability information to teammates and receive reflected messages. The reflected messages refer to the state and capability information of teammates that are transmitted from each teammate to the ad hoc agent. Team members are expected to transmit the corresponding messages back. When planning the next sub-task, the ad hoc will broadcast its intention and utilize the received messages from teammates to complete the planning. The task-relevant state contains the current sub-task and subskill planning results. When the ad hoc agent has a new task-relevant state, specifically when it finishes or switches to new sub-tasks or subskills, it will transmit the information to teammates. The ad hoc agent can also receive the corresponding information from other teammates to do subsequent planning. After execution of each step, the ad hoc agent can reason and judge

Fig. 6.3 The examples of communication messages

whether the currently detected object is misplaced and if it can be the receptacle for other misplaced objects. The key detection information refers to these reasoning results. When the ad hoc agent obtains the new key detection information, it will broadcast this information to other teammates. The examples of the above communication are shown in Fig. 6.3. We have created an interface for different policies of teams to adapt to the ad hoc agent. This interface extracts capability and state information from the message format of teammates' models, accommodating their varying action policies.

6.3.3 Visual Semantic Perception

The visual semantic perception module detects the objects in the current view of the ad hoc agent, generates the room-object-aware semantic map, and obtains the spatial relationships among all detected objects. At each step, the ad hoc agent generates the semantic point clouds based on the obtained RGB-D image, instance segmentation result, and the current pose. The type of room where the agent is located is reasoned according to the categories of currently detected objects. The semantic point clouds are voxelized and flattened to generate a top-down object-room-aware semantic map, which is utilized for exploration and navigation. The hierarchical spatial scene graph is constructed based on the coordinates of each detected object and receptacle as well as the predicted room location, which is utilized to plan the subsequent sub-task and subskill. The ad hoc agent updates the semantic map and scene graph with new detection information at each step.

6.3.4 Dynamic Sub-task Planner with IRoT

The LLM is employed as the dynamic sub-task planner to predict the subsequent sub-task for the ad hoc agent. This sub-task is structured to encapsulate the agent's overall intention within a specific time frame. The LLM is utilized for planning due to its ability to analyze and rationalize the sub-tasks that need to be executed based on teammates' capabilities and state information. It also offers flexibility in collaborating with teammates possessing varied skill sets. When joining and collaborating with the original team, the ad hoc agent is likely to have multiple options of sub-tasks to execute when planning, and performing different sub-tasks may have different effects on the original team. The ad hoc agent needs to find the most suitable sub-task to make the collaboration more efficient. Due to the limited information the ad hoc agent obtains when joining the team, the communication messages from teammates are beneficial for the ad hoc agent to make decisions. To solve this complex situation in the interactive environment, we propose the new method *Interactive Reflection of Thoughts (IRoT)* on the basis of the LLM, which introduces the communication reflection mechanism to evaluate the generated candidate sub-tasks with obtained communication messages and determine the next sub-task for the ad hoc agent.

In the context of the heterogeneous multi-agent tidying-up task, the sub-tasks are categorized into two types:

- Explore $\langle ID_{room} \rangle$: Explore the room ID_{room} to search for misplaced objects and obtain information of the room.
- Replace $\langle ID_{object} \rangle$ $\langle Type_{receptacle} \rangle$ $\langle Type_{room} \rangle$: Put the detected misplaced object ID_{object} to the reasonable type of receptacle $Type_{receptacle}$ in the reasonable type of room $Type_{room}$.
- Stop: Stop at the current location.

This approach contains three stages, which are demonstrated in Fig. 6.4. The first stage is *Generation of Candidate Sub-tasks*, in which the LLM generates n_{irot} candidate sub-tasks and the corresponding reasoning analysis with the provided prompts. The second stage is *Evaluation and Ranking*, in which the LLM evaluates the generated sub-tasks and ranks them based on the execution rationality with the candidate sub-tasks and evaluation prompts as input. The third stage is *Interactive Reflection*. In this stage, ranked sub-tasks are sequentially selected and shared with teammates through communication as the ad hoc agent's intention. The LLM judges the feasibility of the selected sub-task based on the received feedback from teammates. If the selected one is judged to be unreasonable, the next sub-task is

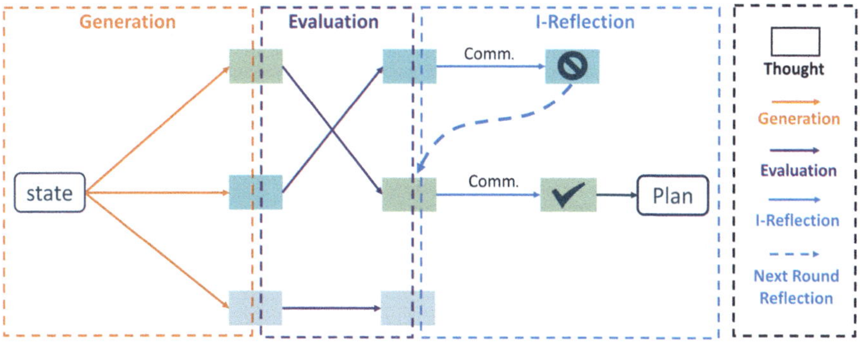

Fig. 6.4 The structure and the planning process of IRoT

selected for another round of this stage. If the selected one is reasonable, the selected sub-task is the planning results of the sub-task planner. The key prompts of the IRoT are shown in Fig. 6.5, and we show the output example of the dynamic sub-task planner with IRoT in our proposed framework and the detailed analysis for each step in Fig. 6.6.

Each rectangle box represents a *thought*, which is an intermediate sub-task plan toward problem solving. Different colors of rectangle boxes signify distinct generated sub-tasks. Lines in different colors represent three stages in the IRoT model. Dotted lines indicate that the chosen sub-task is evaluated to be infeasible, and the next chosen sub-task is evaluated with the communication messages. From left to right, n_{irot} sub-task plans are initially generated in the first stage *Generation of Candidate Sub-tasks*. Then these sub-tasks are evaluated and ranked in the second stage *Evaluation & Ranking*. Finally, IRoT re-judges whether each ranked sub-task is feasible in sequence based on the communication information with teammates in the third stage *Interactive Reflection*. As shown in this figure, the first selected sub-task (*Rank 1*) is evaluated infeasible, and the second selected one (*Rank 2*) is evaluated reasonable to execute. IRoT finally selects the second sub-task to execute.

6.3.4.1 Generation of Candidate Sub-tasks

We first generate n_{irot} candidate sub-tasks with the Chain-of-Thought (CoT) [32] method of the LLM. The state, capability, communication, and memory information are provided to the LLM. It should be noted that all agents obtain partial observations in the environment, which means they can only obtain information about detected objects from their first-person views but cannot get information about all objects in the scene. The action strategies and observation information of other teammates are unknown to the ad hoc agent. The capabilities and executed sub-tasks of each teammate can be extracted from the communication messages. We give several examples of generation results in the prompt to let the LLM give n_{irot} different candidate sub-tasks while avoiding generating duplicated sub-tasks.

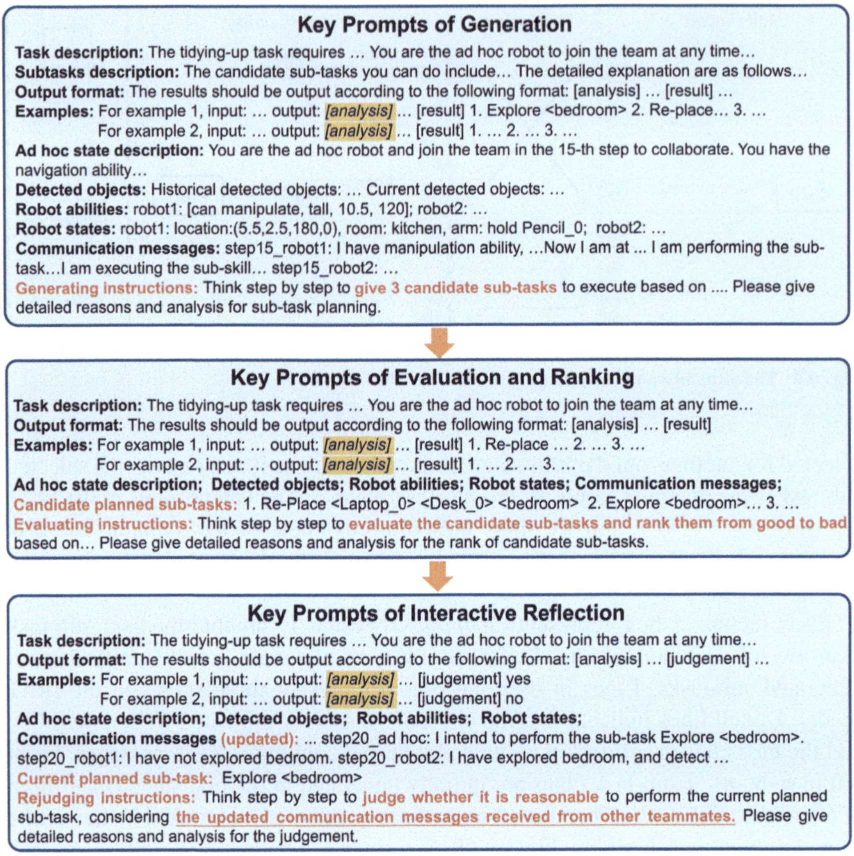

Fig. 6.5 Key prompts of the dynamic sub-task planner with IRoT

6.3.4.2 Evaluation and Ranking

We utilize the LLM to evaluate and rank the n_{irot} generated candidate sub-tasks with the CoT method. We provide the state, capability, communication, and memory information to the LLM. The LLM takes this information and in-context examples to reason the possible future states after executing different candidate sub-tasks and evaluate the effect on the collaboration with teammates. In this planning process, the LLM considers more details, including the capabilities of the ad hoc agent and teammates, the process of the task execution, and the distance to the detected misplaced objects to rank the cooperation effect of candidate sub-tasks. It outputs the ranked results of n_{irot} candidate sub-tasks according to their rationality in cooperation with teammates.

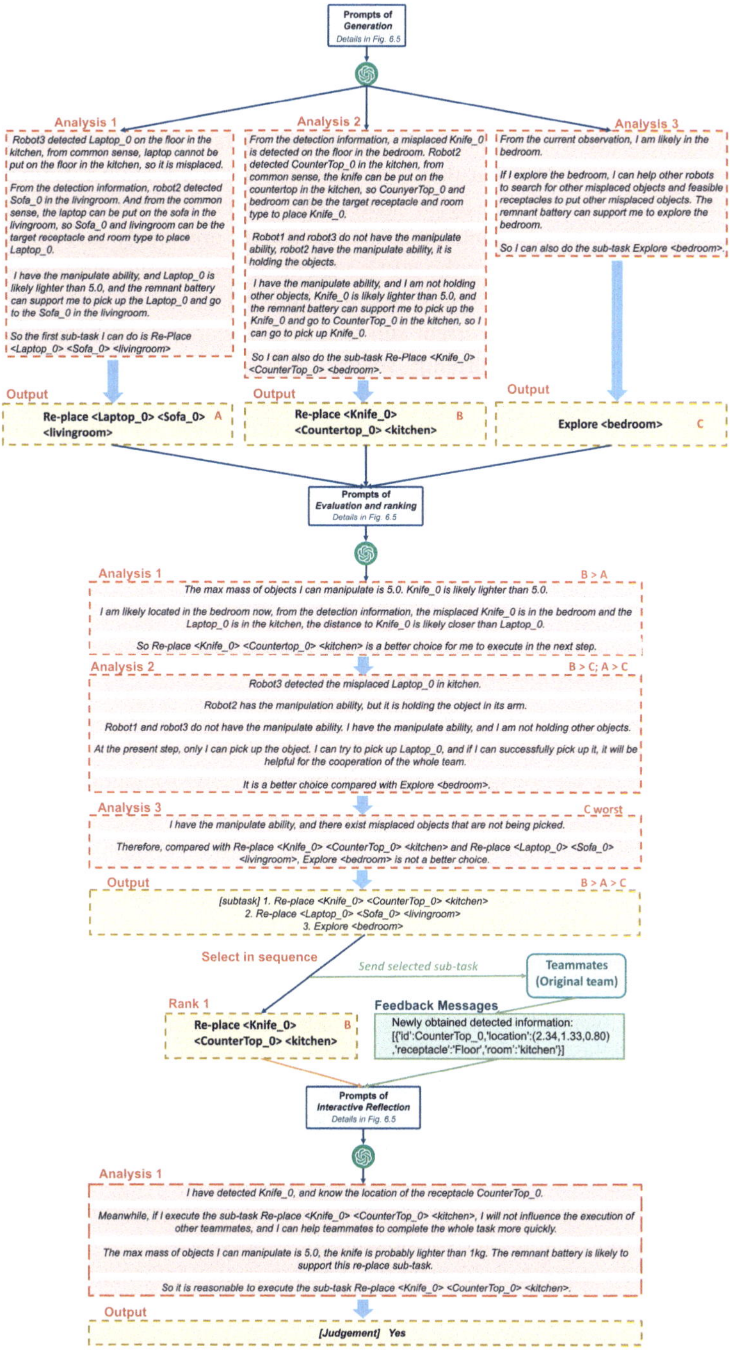

Fig. 6.6 One example of analysis output of the dynamic sub-task planner with IRoT

6.3.4.3 Interactive Reflection

We choose one candidate sub-task from the ranked list of candidate sub-tasks in sequence. The ad hoc agent broadcasts the selected sub-task to other teammates through communication messages. Because the ad hoc agent is likely to join the team after several steps, it does not have the historical state information of teammates. Once the ad hoc agent shares the chosen sub-task, teammates are expected to give responses containing information on previously detected objects or explored rooms. This assists the ad hoc agent in determining the feasibility of the selected sub-task. The LLM uses updated communication and state information to assess the viability of the current sub-task of collaborating with teammates. If the evaluation result of the currently selected sub-task is reasonable, the ad hoc agent will perform this sub-task and make subsequent subskill planning based on this sub-task. If the evaluation result is unreasonable, the method will select the next generated sub-task from the ranked list and evaluate it in the same way as above. If the evaluation results of all n_{irot} candidate sub-tasks are not reasonable, the method will replan the sub-task from the first step with the newly obtained state information of the task.

6.3.5 Dynamic Subskill Planner

The dynamic subskill planner generates the next subskill with LLM based on the CoT method for the ad hoc agent. The execution of a sub-task can be decomposed into a list of subskills within a certain time sequence. The candidate subskills include the following subskills:

- GoToObject $\langle objectID\rangle$: Navigate to the vicinity of the detected object $objectID$ in the current view.
- GoToPoint $\langle \Delta x, \Delta y\rangle$: Navigate to the relative position $(\Delta x, \Delta y)$ from the current position, and Δx, Δy represent the distance to the target point in the x and y axis, respectively, within a certain distance constraint. This subskill corresponds to the embodied PointNav task.
- GoToRoom $\langle roomID\rangle$: Navigate to the location of the known room $roomID$.
- PickupObject $\langle objectID\rangle$: Pick up the detected object $objectID$ when the agent is located near it.
- PutObject $\langle objectID\rangle$ $\langle receptacleID\rangle$: Place the object $\langle objectID\rangle$ being picked up to the target receptacle $\langle receptacleID\rangle$. The execution of this subskill requires that the agent to pick up the corresponding object.
- Communicate $\langle message\rangle$: Convey the communication information $\langle message\rangle$ to other teammates.
- Explore: Explore the specific area to obtain more information about the scene.
- Stop: Stop at the current location.

The subskills are the meta skills that can be executed by the controller. We provide the state, capability, communication, and memory information, as well as the in-context examples in the prompts to the LLM. Specifically, the current sub-task and subskill plan of each teammate can be obtained from the state and memory information. The key prompts of planning subskills are shown in Fig. 6.7, and we show the output example of the dynamic subskill planner in Fig. 6.8, which demonstrates the analysis and reasoning process of the planner.

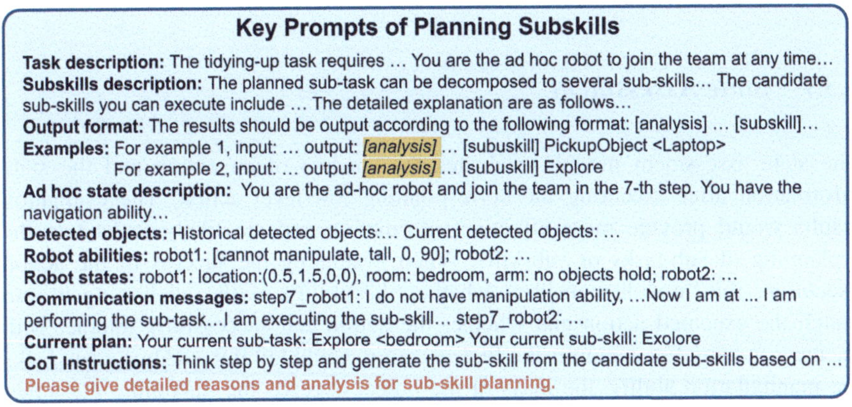

Fig. 6.7 Key prompts of the dynamic subskill planner

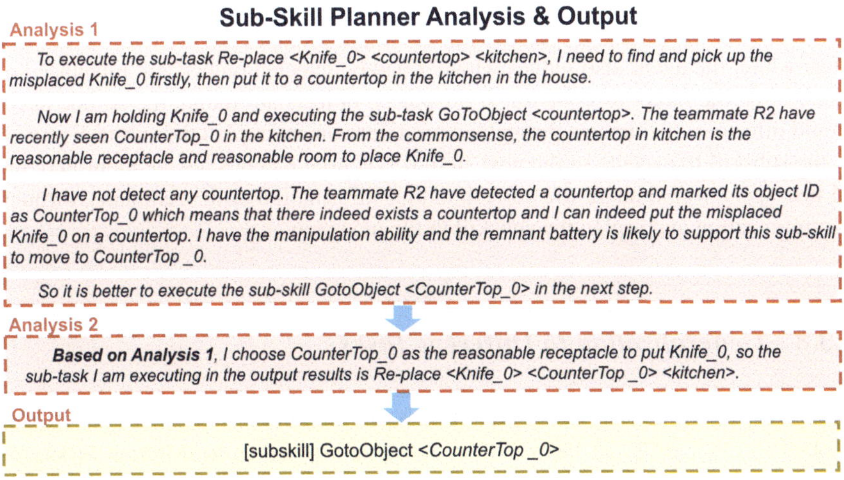

Fig. 6.8 Example of the analysis and output of the dynamic subskill planner

6.3.6 Subskill Execution

The subskill execution module generates the low-level actions to execute the subskill. This module executes the planned subskill and updates the ever-changing state information, informing both the sub-task and subskill planners of necessary adjustments. We can obtain the execution policies for each subskill in advance. The execution policies can be generated through rule-based methods or trained by imitation learning (IL) and reinforcement learning (RL).

6.3.7 State Assessment

The state assessment module evaluates the success of an action and the state information after executing the corresponding low-level action. The evaluation results would provide more effective information and are likely to lead to the replanning of sub-tasks or subskills. When evaluating the success of the action execution, this module considers whether the ability of the ad hoc agent can match the executed action and whether the action can successfully interact with the environment. For example, the *Pickup* action would fail if the ad hoc agent lacks the manipulation ability, the mass of the object exceeds the maximum mass that the agent can pick, or the agent is far away from the target object when executing the *Pickup* action. If the agent encounters obstacles when performing the navigation action, the action execution would also fail. If the action execution fails, this module would identify the failure reason and provide feedback to the planner and the action execution module to regenerate the reasonable subskills and feasible actions. Meanwhile, the agent also judges whether newly detected objects are misplaced or likely to be the target receptacles. If misplaced objects are found, the ad hoc agent would inform its teammates through communication and replan the sub-task as well as the subskill to execute in the next step. If potential receptacles are identified, the ad hoc agent would share the corresponding information with teammates through communication and replan the subskill in the next step.

6.3.8 Generalization to Different Tasks

The proposed framework with the LLM-based hierarchical planner can also be applied to other different tasks. When applying our model to different tasks, the structure of the model remains unchanged. The dynamic sub-task planner with IRoT and the dynamic subskill planner are still utilized to generate the next sub-task and the subskill to execute. To adapt to different tasks, it is required to design corresponding reasonable candidate sub-tasks and subskills for the task, update the description of the new task, modify the descriptions of sub-tasks and subskills in the corresponding prompts, and keep the structure as well as other contents in the prompts of the sub-task and subskill planner unchanged.

As for the sub-task planner, the whole process of IRoT is kept unchanged, and the candidate sub-tasks are designed according to the nature of the corresponding tasks. Then it is necessary to update the description of the new task and modify the description of candidate sub-tasks in the prompts of the sub-task planner as shown in Fig. 6.5 to the description corresponding to the newly defined sub-tasks. The other contents in the prompts of the sub-task planner can be kept unchanged. The modified model can generate corresponding sub-tasks in new tasks. Meanwhile, as for more complex tasks, if all possible candidate sub-tasks cannot be completely predefined, we can remove the sub-task descriptions in the prompts without setting the sub-task types in advance and let the LLM directly generate reasonable sub-tasks according to current information. Then the generated sub-task is input to the subskill planner to generate the corresponding subskills. The candidate subskills are predefined, and the output of the subskill planner must come from the predefined types of subskills. The structure and other contents in the prompts of the sub-task planner can be kept unchanged.

The subskill planner generates the basic skill for the agent to execute according to the current sub-task. In common indoor tasks such as visual navigation [34], visual rearrangement [33], and ALFRED [26], the manipulation and interaction on objects can be eventually decomposed into the subskills of navigation, picking, and placing, so the seven candidate subskills proposed in our model can cover basic skills in common tasks. When applying our model to different tasks, the task description needs to be updated, and other contents in the prompts of the subskill planner can remain unchanged. If more complex tasks require other subskills to be executed, it is required to design new subskills and provide the low-level action strategy under the corresponding subskills. Then the subskill descriptions in the prompts as shown in Fig. 6.7 need to be modified to the descriptions of newly defined subskills. Other contents in the prompts can be kept unchanged in the subskill planner. The subskill planner can generate the next subskill to execute from the candidate subskills.

To apply our model to other tasks, we only need to design the candidate sub-tasks and subskills according to the properties of the new task, update the description of the new task, and modify descriptions of the sub-task as well as the subskill in the prompts of the sub-task planner and the subskill planner.

6.4 Simulation Experiments

6.4.1 Dataset

As far as we know, there is not any existing benchmark dataset that can evaluate heterogeneous ad hoc teamwork. Considering the tidying-up is an ideal task to evaluate the performance of heterogeneous ad hoc teamwork, we build a benchmark testing dataset for the heterogeneous ad hoc teamwork on the tidying-up task with the ProcTHOR-10K [6] dataset.

Scenario Generation We randomly select 90 houses from the testing houses in ProcTHOR-10K for evaluation experiments, among which every 10 houses are with 1–8, and 10 rooms, respectively (there is no house with nine rooms in ProcTHOR-10K). We generate five tidying-up scenarios for each house, where k pickupable objects are randomly selected and misplaced to unreasonable receptacles or dropped in a random location in the house ($k = 1, 2, 3, 4, 5$).

Heterogeneous Teammates Generation The characteristics of the teammate are described by a tuple of $(\alpha_{nav}, \alpha_{manip}, h, m, step)$ where α_{nav} and α_{manip} indicate the navigation ability and manipulation ability of the agent, h is the height of the agent, m denotes the payload which means the maximum weight of the object that can be picked up by the agent, and $step$ is the max steps that the agent's battery can support. For each tidying-up scenario, we consider situations of having 3, 4, and 5 teammates, respectively, and we randomly generate the values for each teammate's characteristics. It is guaranteed that all generated agents have the navigation ability, and at least one agent has manipulation ability in order to accomplish the tidying-up task. Additionally, the initial positions of each teammate are randomly generated from the reachable positions in the corresponding task.

Ad hoc Agent Generation Similar to the process of generating heterogeneous teammates. We generate one ad hoc agent for each tidying-up task. It is noted that the ad hoc agent does not have to have the manipulation ability. Its initial position is also randomly generated.

Finally, we obtain 1350 tidying-up tasks for the heterogeneous ad hoc teamwork. We divide these tasks into three difficulty categories: *Easy*, *Medium*, and *Difficult*, in which houses with 1–3 rooms are *Easy* tasks, houses with 4–6 rooms are *Medium* tasks, and houses with 7–10 rooms are *Difficult* tasks. The heterogeneous agent and scene settings are shown in Fig. 6.9. The process of generating the data can be applied and extended to the whole testing set in ProcTHOR-10K.

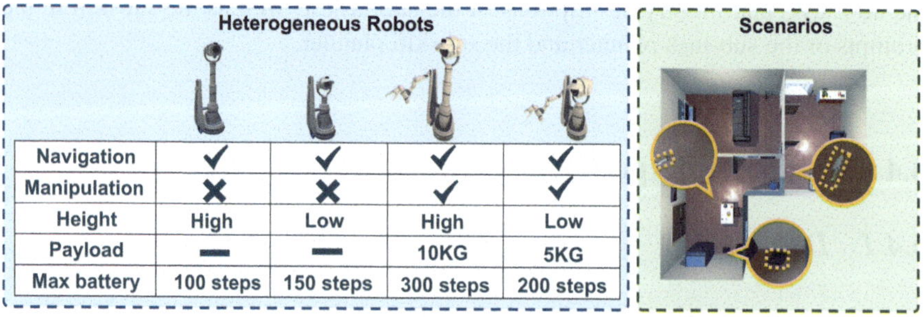

Fig. 6.9 The heterogeneous agents and scene settings in the benchmark dataset

It is worth noting that the tidying-up task is different from the rearrangement task. In the rearrangement task, the agent only needs to replace objects according to the given initial state, while in the tidying-up task, the agent is required to infer the reasonable locations of each object by itself without extra guidance and replace misplaced objects to reasonable locations, which puts forward higher requirements for scene understanding and commonsense reasoning.

6.4.2 *Heterogeneous Teammates Policy*

Before the ad hoc agent joins the team, the original heterogeneous teammates collaborate with each other with a certain policy. The proposed ad hoc teamwork framework can adapt to different heterogeneous teammates no matter what policy they employ. In the experiment, we consider the following different common policies for the original heterogeneous teammates' collaboration.

6.4.2.1 Heuristic-Based Policy

A heuristic-based policy is designed for the tidying-up task. When a teammate with manipulation ability detects the misplaced object, it will directly choose to pick up the detected misplaced object and place it in a reasonable location. Teammates without manipulation ability will continue exploring the scene. Specifically, if no misplaced object is detected by teammates, they will collaboratively explore the whole house based on the frontier-based exploration method in the multi-agent setting. If someone detects a misplaced object, then the agent that has the manipulation ability and is closest to the misplaced objects based on the predicted location information will choose to perform the sub-task to replace the misplaced object to the reasonable receptacle. Other teammates will continue exploring the house. Whenever a misplaced object is detected, the sub-task of replacing the newly detected misplaced object will be assigned to the teammate that satisfies the above conditions and is not performing the replace sub-task. If all teammates with the manipulation ability have been assigned the replace sub-task, it is required to wait for someone to complete the executing replace sub-task and then switch to the new replace sub-task.

6.4.2.2 Learning-Based Policy

We utilize imitation learning to train the learning-based teammate's policy to predict the subskill for the teammate to execute. The learning-based model includes the visual semantic perception, subskill decision, and the action execution module. The visual semantic perception is similar to that in our proposed model, which obtains the semantic map and the hierarchical scene graph. In the learning-based method, the visual semantic perception module also includes a misplaced object detector consisting of two linear layers. It takes the detected object, its current receptacle, and the current room type as the input and outputs whether the object is misplaced. We train it in the training scenes of ProcTHOR-10k with the ground truth of the object placement constraints in ProcTHOR in advance. The subskill decision module is trained to predict the subskill for the agent. It encodes the abilities of the agent, the current location, and state of the agent, as well as the spatial relationship of detected objects to obtain the detection embedding and fuse these embeddings to get the overall state feature. Then it fuses its state feature with others' state features and utilizes the fused feature to predict the type of the next subskill. The candidate type of the predicted subskill for this learning method includes *Explore*, *PickupObject*, *PutObject*, and *Stop*. The action execution module executes the predicted subskill with the rule-based strategy. We also construct a reasonable receptacle predictor based on the neural network linear layers. It takes the type of the misplaced object as input and outputs the chosen reasonable type of the receptacle and room from the candidate list in ProcTHOR. The reasonable receptacle predictor is also pretrained with the metadata of the object placement constraints in ProcTHOR. When the predicted sub skill type is *PickupObject*, if an object is detected to be misplaced, then the agent will choose to pick up the corresponding object with the action execution module; otherwise, the agent will continue exploring the scene. If the predicted subskill type is *PutObject*, the action execution module will firstly predict the reasonable location with the reasonable receptacle predictor and then put the hold object to the predicted location. We utilize the metadata of the task and scene to generate the expert demonstration for the predicted subskill for each agent. We train the subskill decision module with imitation learning by the behavior cloning method.

6.4.2.3 LLM-Based Policy

The structure of the teammates' model based on the LLM consists of the visual semantic perception module, negotiation-based sub-task planner, subskill planner, subskill execution, and the communication module.

Visual Semantic Perception The visual perception module is utilized to generate the scene graph and semantic map in the same way as the visual semantic perception module in the proposed ad hoc teamwork framework.

Negotiation-Based Sub-task Planner The negotiation-based sub-task planner generates the next sub-task based on the negotiation dialog among teammates with the LLM, which includes *Explore* $\langle ID_{room} \rangle$ and *Replace* $\langle ID_{object} \rangle$ $\langle Type_{receptacle} \rangle$ $\langle Type_{room} \rangle$ in the multi-agent tidying-up task. Due to the different abilities of heterogeneous agents, each agent needs to decide the next sub-task to execute while carefully considering their skills and current states when the misplaced objects are detected. The predicted sub-tasks of each agent need to meet the requirements of their own abilities and avoid planning conflicts among agents. We utilize the reasoning and planning abilities of the LLMs to complete sub-task planning with the negotiation dialog of agents. When the task starts, each agent communicates its ability characteristics to other teammates and plans the next sub-task based on the CoT approach with the information of teammates' abilities and observations in sequence. Each teammate's abilities, the information of currently detected objects, the historical planning results, and the current predicted sub-task by other teammates who have already finished planning are provided to the LLM-based planner to generate the next sub-task for the corresponding agent. The planner first generates the answer whether it agrees with the already finished planning results by previous agents. If the agent agrees to the finished planning results, it then outputs its planned sub-task to perform. Otherwise, the previous agents will negotiate and replan the sub-task again with this feedback information.

Dynamic Subskill Planner The dynamic subskill planner leverages the LLM to predict the next subskill with the CoT method, considering their abilities, current and historical states, communication messages as well as the predicted sub-tasks. Each teammate generates the next subskill to execute in a similar way as the dynamic subskill in the proposed ad hoc teamwork model.

Subskill Execution The subskill execution module executes the predicted subskill and updates the newly detected information. We generate policies for each subskill in advance and utilize the rule-based methods based on the frontier-based approach to generate low-level actions.

Communication Each teammate utilizes the communication module to transmit the key information. At the beginning of the task, each agent broadcasts its ability and detection information through communication messages. Then teammates utilize the extracted ability and state information to plan the next sub-tasks and subskills. Similar to the communication module in the proposed ad hoc teamwork framework, teammates transmit key information of detected misplaced objects and candidate receptacles to others, as well as broadcast the state information when finishing or switching to new sub-tasks and subskills.

Adaptive Adjustment to Ad Hoc Agent In our ad hoc teamwork task, we consider whether the strategies of teammates adapt to the policy of the ad hoc agent after the ad hoc agent joins the teamwork. Therefore, we also consider utilizing the extra adaptive prompts when planning to explicitly make teammates ponder the situation in which an existing agent is broken or a new ad hoc agent is added to the team.

6.4.3 Ad Hoc Agent Policy

For the ad hoc agent, we compare the proposed method with several baselines as follows:

1. **Random Policy**: The ad hoc agent randomly chooses one action from its action space to execute at each step.
2. **Heuristic-Based Policy**: Similar to the heuristic-based for the original heterogeneous teammates, for the ad hoc agent with manipulation ability, if a misplaced object is detected, it will directly pick it up and place it in a reasonable location. For the ad hoc agent without manipulation ability, it will directly explore the scene from its current location.
3. **PLASTIC Policy**: We train the ad hoc policy with several predefined teams based on the existing common method of *PLASTIC* [3], and the predicted subskill is executed with imitation learning. The ad hoc agent directly uses the learned policy to collaborate with unknown heterogeneous teammates.
4. **Naive-LLM Policy**: We utilize the LLM to directly predict the sub-task for the ad hoc agent without specific prompt engineering for thinking and reasoning. The subskill planning and execution are the same as the proposed policy.
5. **GPL-LLM Policy**: Most of the existing learning methods for ad hoc teamwork are suitable for a simple 2D environment, and almost no existing method can be directly applied to a 3D environment because of the substantially increased dimensions of visual information and the complex characteristics of partial observation and interaction with the environment. The latest Graph-Based Policy Learning method [21] is also difficult to be directly transferred to our task, but we also hope to verify the effect of this idea in our ad hoc task, so we combine the idea with the LLM to form the baseline *GPL-LLM*. We utilize the LLM to first predict the next sub-task and subskill of all teammates based on the current states and capabilities of all agents. We let the LLM generate candidate sub-tasks that can be executed by the ad hoc agent and give the evaluation score for each generated sub-task according to the predicted plans of teammates. The final sub-task for the ad hoc agent to execute is generated according to the evaluation score.
6. **CoT-LLM Policy**: We utilize the CoT prompts to predict sub-tasks for the ad hoc agent without the evaluation and communication reflection mechanism. The subskill planning and execution are the same as our proposed policy.

6.4.4 Implementation Details

We call our method based on IRoT as the ***IRoT-LLM*** model. All the LLM-based baselines and our *IRoT-LLM* model utilize the *gpt-4-1106-preview* to complete the planning process. The parameter *Temperature* is set to be 0.3 in our experiments. The maximum limited number of tokens in the prompt is 128,000. If the maximum limit is exceeded in tasks, we would move out some of the earliest information from the historical communication messages in the prompt to adapt to the maximum limit.

6.4.5 Evaluation Metrics

The effectiveness of an ad hoc policy can be quantified by the difference in performance between the original heterogeneous team and the new team after the ad hoc agent joins. We denote N_{epi} as the number of tasks. In the context of our experiment, we utilize the evaluation metrics as follows:

1. **%SR**: The success rate of tasks, which is calculated as follows:

$$\%SR = \frac{1}{N_{epi}} \sum_{i=1}^{N_{epi}} R_i,$$

   where $R_i = 1$ if all misplaced objects are replaced in reasonable locations, otherwise $R_i = 0$. It denotes the strict success rate.
2. **%PS**: The partial success rate of tasks, which is calculated as follows:

$$\%PS = \frac{1}{N_{epi}} \sum_{i=1}^{N_{epi}} \frac{K_i^{suc}}{K_i},$$

   where K_i is the number of misplaced objects in the i-th task, and K_i^{suc} is the number of misplaced objects that are replaced to reasonable locations. It reflects the proportion of misplaced objects that are successfully tidied.
3. **#TS**: The average temporal steps of task completion, which is calculated as follows:

$$\#TS = \frac{1}{N_{epi}} \sum_{i=1}^{N_{epi}} TS_i,$$

   where TS_i is the number of temporal steps when completing the i-th task, the value equals the maximum steps among all agents' steps when they stop. If agents do not successfully complete the task, TS_i is the maximum number of steps. It evaluates the temporal efficiency of completing the task. The fewer $\#TS$, the higher the efficiency.
4. **#AS**: The total action steps of all agents in tasks, which are calculated as follows:

$$\#AS = \frac{1}{N_{epi}} \sum_{i=1}^{N_{epi}} AS_i,$$

   where AS_i is the sum of action steps of all agents when they stop in the i-th task. If agents do not successfully complete the task, the action step of each agent is the maximum number of steps. It can evaluate the total energy consumption after the ad hoc agent joins the team. The fewer $\#AS$, the less energy consumption of the collaboration.

5. **%SE**: Success rate weighted by efficiency improvement, which is calculated as follows:

$$SE = \frac{1}{N_{epi}} \sum_{i=1}^{N_{epi}} \frac{max(0, TS_i^{team} - TS_i^{adhoc})}{TS_i^{team}} R_i^{adhoc},$$

where TS_i^{team} is the number of temporal steps when only teammates completing the i-th task, and TS_i^{adhoc} is the number of temporal steps when the ad hoc agent joins the team to complete the i-th task. It can evaluate the efficiency improvement of the ad hoc agent collaboration compared with the existing team.

6.4.6 Quantitative Performance

We evaluate the performance of our ad hoc framework collaborating with different policies of teammates at different times; the results of joining the team at $t_0 = 0$, $t_0 = 50$ $t_0 = 100$ are illustrated in Table 6.1, 6.2, and 6.3, respectively. We call our method based on IRoT as the **IRoT-LLM** policy. We calculate the value improvement percentage for the metrics that are better with the higher value ($\%SR$, $\%PS$), calculate the decline percentage for the metrics that are better with the lower value ($\#TS$, $\#AS$), and put this value in brackets. The performance shows the influence of different ad hoc methods on the original team when the ad hoc agent joins the team and collaborates with teammates. From the results, we obtain the following findings:

(1) **Our framework can obviously improve the success rate and efficiency of the original team of different policies compared with no ad hoc agent joining the team.** Our method improves $\%SR$, $\%PS$, $\%SE$ and reduces $\#TS$, $\#AS$ when collaborating with the heuristic-based, learning-based, and LLM-based teams, which demonstrates that our policy can adapt to different policies of teams. The improvements on the original team of our policy differ in collaborating with a team of different policies, and the improvement can be larger when the team policy itself is worse. The performance of different ad hoc policies in $\%SR$ collaborating with the LLM-based team at different times is also illustrated in Fig. 6.10a.

(2) **Our framework can improve the performance of the original teams of different policies when joining them at different times** of $t_0 = 0$, $t_0 = 50$, and $t_0 = 100$. It is noticed in Fig. 6.10a, the later the ad hoc agent joins the original team, the less improvement of the ad hoc agent on this team. This is because teammates have already cooperated with each other and completed more parts of the task when the ad hoc agent joins the team later. The results

Table 6.1 Quantitative results in ad hoc teamwork with different team policies when $t_0 = 0$

Team	ad hoc	$\%SR(\uparrow)$	$\%PS(\uparrow)$	$\#TS(\downarrow)$	$\#AS(\downarrow)$	$\%SE(\uparrow)$
Heuristic-based	No ad hoc	15.3 (-)	21.7 (-)	467.3 (-)	1863.2 (-)	–
	Random	15.3 (0)	21.7 (0)	467.3 (0)	2213 (-18.8)	0
	Heuristic	18.1 (18.3)	27.4 (26.3)	435.6 (6.8)	2049.3 (-10.0)	1.2
	PLASTIC	15.6 (2.0)	22.3 (2.8)	456.1 (2.4)	2165.9 (-16.2)	0.4
	Naive-LLM	18.3 (19.6)	27.9 (28.6)	436.2 (6.7)	2060.1 (-10.6)	1.2
	GPL-LLM	18.5 (20.9)	27.9 (28.6)	433.9 (7.1)	2046.1 (-9.8)	1.3
	CoT-LLM	19.9 (30.0)	28.5 (31.3)	421.3 (9.8)	1987.6 (-6.7)	2.0
	IRoT-LLM	**21.0 (37.3)**	**30.7 (41.5)**	**410.8 (12.1)**	**1922.3 (-3.2)**	**2.6**
Learning-based	No ad hoc	13.5 (-)	20.8 (-)	473.8 (-)	1890.2 (-)	–
	Random	13.5 (0)	20.8 (0)	473.8 (0)	2212.9 (-17.1)	0
	Heuristic	15.6 (15.6)	25.1 (20.7)	453.1 (4.4)	2133.1 (-12.9)	0.6
	PLASTIC	13.8 (2.2)	22.6 (8.7)	463.8 (2.1)	2192.1 (-16.0)	0.3
	Naive-LLM	15.7 (16.3)	25.5 (22.6)	449.2 (5.2)	2111.9 (-11.7)	0.8
	GPL-LLM	15.9 (17.8)	25.9 (24.5)	450.3 (5.0)	2115.3 (-11.9)	0.9
	CoT-LLM	16.9 (25.2)	26.4 (26.9)	440.1 (7.1)	2065.6 (-9.3)	1.2
	IRoT-LLM	**18.1 (34.1)**	**27.3 (31.3)**	**431.2 (9.0)**	**2010.2 (-6.3)**	**1.7**
LLM-based	No ad hoc	21.1 (-)	27.2 (-)	453.2 (-)	1813.3 (-)	–
	Random	21.1 (0)	27.2 (0)	453.2 (0)	2186.1 (-20.1)	0
	Heuristic	23.1 (9.4)	33.1 (21.7)	417.6 (7.9)	2013.3 (-11.0)	1.8
	PLASTIC	21.5 (1.9)	28.0 (2.9)	440.2 (2.9)	2130.6 (-17.5)	0.6
	Naive-LLM	23.3 (10.4)	33.5 (23.2)	418.1 (7.7)	2015.2 (-11.1)	1.8
	GPL-LLM)	23.5 (11.4)	33.7 (23.9)	415.5 (8.3)	2013.5 (-11.0)	1.9
	CoT-LLM	24.9 (18.0)	33.7 (23.9)	406.6 (10.3)	1945.3 (-7.3)	2.5
	IRoT-LLM	**26.3 (24.6)**	**35.1 (29.0)**	**399.3 (11.9)**	**1918.1 (-5.8)**	**3.2**

demonstrate that the ad hoc agent with the proposed policy can still adapt to the team no matter when it joins the team and improve the performance of the team in the subsequent execution of the task.

(3) **Our framework performs better than other ad hoc policies when collaborating with different teams.** The ad hoc policies based on the LLM including *Naive-LLM*, *GPL-LLM*, *CoT-LLM*, and *IRoT-LLM* achieve higher $\%SR$, $\%PS$, and $\%SE$ as well as lower $\#TS$, $\#AS$ than the learning-based baseline *PLASTIC*, since LLM-based methods can deal with unseen complex circumstance more flexibly, and have better generalization ability. *Naive-LLM* performs similarly to *heuristic-based* policy, which to some extent shows that planning in the embodied environment is more difficult compared with a simple static planning problem. *IRoT-LLM* performs the best among all LLM-based ad hoc policies, demonstrating the effectiveness of the IRoT method with the communication messages. The performance of *GPL-LLM* shows that although most of the existing learning methods for ad hoc teamwork are difficult to be directly transferred to our 3D task, combining the key idea with the LLM can

Table 6.2 Quantitative results in ad hoc teamwork with different team policies when $t_0 = 50$

Team	ad hoc	$\%SR(\uparrow)$	$\%PS(\uparrow)$	$\#TS(\downarrow)$	$\#AS(\downarrow)$	$\%SE(\uparrow)$
Heuristic-based	No ad hoc	15.3 (-)	21.7 (-)	367.3 (-)	1863.2 (-)	–
	Random	15.3 (0)	21.7 (0)	467.3 (0)	2183.2 (-17.2)	0
	Heuristic	17.4 (13.7)	26.7 (23.0)	439.2 (6.0)	2001.3 (-7.4)	1.1
	PLASTIC	15.5 (1.3)	22.1 (1.8)	461.3 (1.3)	2132.9 (-14.5)	0.3
	Naive-LLM	17.2 (12.4)	27.0 (24.4)	440.8 (5.7)	2006.2 (-7.7)	1.1
	GPL-LLM	17.4 (13.7)	27.2 (25.3)	436.1 (6.7)	1996.3 (-7.1)	1.1
	CoT-LLM	18.6 (21.6)	27.6 (27.2)	423.2 (9.4)	1936.3 (-3.9)	1.8
	IRoT-LLM	**20.3 (32.7)**	**29.5 (35.9)**	**415.8 (11.0)**	1873.1 (-0.5)	2.3
Learning-based	No ad hoc	13.5 (-)	20.8 (-)	473.8 (-)	1890.2	–
	Random	13.5 (0)	20.8 (0)	437.8 (0)	2176.2 (-15.1)	0
	Heuristic	15.3 (13.3)	24.7 (18.8)	458.2 (3.3)	2092.1 (-10.7)	0.6
	PLASTIC	13.8 (2.2)	22.3 (7.2)	465.1 (1.8)	2156.9 (-14.1)	0.3
	Naive-LLM	15.3 (13.3)	25.1 (20.7)	452.1 (4.6)	2052.7 (-8.6)	0.6
	GPL-LLM	15.6 (15.6)	25.2 (21.2)	452.1 (4.6)	2050.9 (-8.5)	0.7
	CoT-LLM	16.1 (19.3)	26.0 (25.0)	445.3 (6.0)	2016.2 (-6.7)	1.0
	IRoT-LLM	**17.7 (31.1)**	**26.8 (28.8)**	**435.9 (8.0)**	**1995.3 (-5.6)**	**1.5**
LLM-based	No ad hoc	21.1 (-)	27.2 (-)	453.2 (-)	1813.1 (-)	–
	Random	21.1 (0)	27.2 (0)	453.2(0)	2133.3 (-17.6)	0
	Heuristic	22.5 (6.6)	32.1 (18.0)	423.3 (6.6)	1988.6 (-9.7)	1.5
	PLASTIC	21.3 (0.9)	27.5 (1.1)	443.7 (2.1)	2081.3 (-14.8)	0.5
	Naive-LLM	22.5 (6.6)	32.6 (19.9)	425.3 (6.2)	1986.1 (-9.5)	1.4
	GPL-LLM	22.7 (7.6)	32.9 (21.0)	423.1 (6.6)	1988.0 (-9.6)	1.5
	CoT-LLM	24.0 (13.7)	33.1 (21.7)	412.1 (9.1)	1891.9 (-4.3)	2.2
	IRoT-LLM	**25.6 (21.3)**	**34.2 (25.7)**	**403.9 (10.9)**	**1859.2 (-2.5)**	**2.6**

still achieve moderate results. When the ad hoc agent joins the original team, it would improve the efficiency of the team but also might bring more energy consumption due to the addition of one more agent. The $\#AS$ of *IRoT-LLM* when joining the team at $t_0 = 100$ is even smaller than the original team, which demonstrates the proposed method can effectively improve the efficiency of the existing team.

Figure 6.10a shows the performance of different ad hoc policies collaborating with the LLM-based teammates when joining at different times. Figure 6.10b demonstrates the performance of our IRoT-LLM policy collaborating with the LLM-based teammates in different categories of tasks. The histogram represents the value of $\%SR$, and the points on the line chart represent the improvement percentage compared with teammates with no ad hoc agent. Figure 6.10c shows the performance of the ad hoc policies when the ad hoc agent has different abilities.

More specifically, we separately compare the results of the proposed policy in three difficulty categories when collaborating with the LLM-based teammates, which is shown in Fig. 6.10b. The comparison performance illustrates that the

Table 6.3 Quantitative results in ad hoc teamwork with different team policies when $t_0 = 100$

Team	ad hoc	$\%SR(\uparrow)$	$\%PS(\uparrow)$	$\#TS(\downarrow)$	$\#AS(\downarrow)$	$\%SE(\uparrow)$
Heuristic-based	No ad hoc	15.3 (-)	21.7 (-)	467.3 (-)	1863.2 (-)	–
	Random	15.3 (0)	21.7 (0)	467.3 (0)	2152.9 (-15.6)	0
	Heuristic	16.6 (8.5)	25.9 (19.4)	442.1 (5.4)	1967.3 (-5.6)	1.0
	PLASTIC	15.4 (0.7)	21.9 (0.9)	465.2 (0.4)	2107.2 (-13.1)	0.2
	Naive-LLM	16.9 (10.5)	26.3 (21.2)	443.3 (5.1)	1955.9 (-5.0)	0.9
	GPL-LLM	17.1 (11.8)	26.3 (21.2)	438.6 (6.1)	1945.6 (-4.5)	1.0
	CoT-LLM	18.0 (17.6)	27.3 (25.8)	427.9 (8.4)	1877.3 (-0.8)	1.5
	IRoT-LLM	**19.7 (28.8)**	**27.8 (28.1)**	**420.1 (10.1)**	**1822.8 (2.1)**	**2.0**
Learning-based	No ad hoc	13.5 (-)	20.8 (-)	473.8 (-)	1890.2 (-)	–
	Random	13.5 (0)	20.8 (0)	473.8 (0)	2152.9 (-13.9)	0
	Heuristic	15.0 (11.1)	24.1 (15.9)	461.2 (2.7)	2061.2 (-9.0)	0.4
	PLASTIC	13.7 (1.5)	22.1 (6.3)	469.9 (0.8)	2133.8 (-12.9)	0.1
	Naive-LLM	15.1 (11.9)	24.8 (19.2)	455.3 (3.9)	2003.8 (-6.0)	0.6
	GPL-LLM	15.3 (13.3)	24.7 (18.8)	453.1 (4.4)	1995.6 (-5.6)	0.6
	CoT-LLM	15.7 (16.3)	25.5 (22.6)	449.2 (2.5)	1973.9 (-4.4)	0.8
	IRoT-LLM	**17.2 (27.4)**	**26.1 (25.5)**	**439.7 (7.2)**	**1948.2 (-3.1)**	**1.2**
LLM-based	No ad hoc	21.1 (-)	27.2 (-)	453.2 (-)	1813.3 (-)	–
	Random	21.1 (0)	27.2 (0)	453.2 (0)	2106.3 (-16.2)	0
	Heuristic	21.8 (3.3)	31.9 (17.3)	427.6 (5.6)	1932.1 (-6.6)	1.3
	PLASTIC	21.3 (0.9)	27.3 (0.4)	449.1 (0.9)	2067.3 (-14.0)	0.2
	Naive-LLM	22.0 (4.3)	31.8 (16.9)	428.1 (5.5)	1933.9 (-6.7)	1.3
	GPL-LLM	22.0 (4.3)	31.9 (17.3)	426.6 (5.9)	1930.9 (-6.5)	1.3
	CoT-LLM	23.2 (10.0)	32.3 (18.8)	415.3 (8.4)	1840.2 (-1.5)	2.0
	IRoT-LLM	**25.0 (18.5)**	**33.1 (21.7)**	**407.5 (10.1)**	**1803.1 (0.6)**	**2.6**

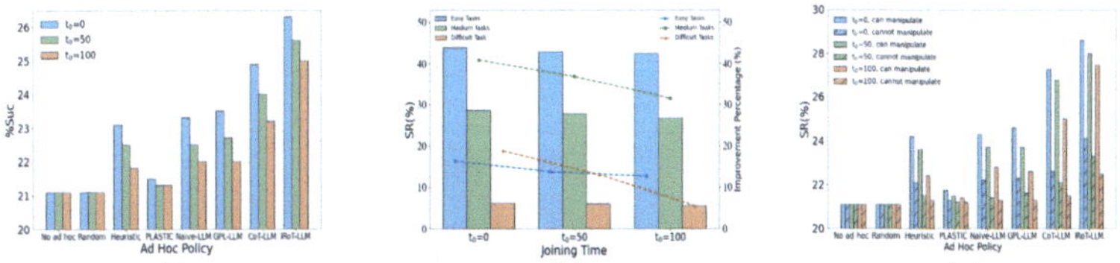

(a) (b) (c)

Fig. 6.10 The performance of different ad hoc policies. (**a**) Performance in $\%SR$ of different ad hoc policies. (**b**) Performance in different task categories. (**c**) Performance in different ad hoc ability

proposed model achieved better improvement after the joining of the ad hoc agent in the *Medium* tasks since the moderate number of rooms provides a good chance for the newly joined ad hoc agent to effectively help the original team and avoid conflicts with the original team cooperation policies. In *Easy* tasks, the existing team can already achieve better performance, so the performance improvement is not obvious. In *Difficult* tasks, the large number of rooms makes it more difficult to find and tidy all misplaced objects, so the improvement of ad hoc teamwork is also limited. Meanwhile, we further compare the influence on the existing LLM-based team and whether the joined IRoT-based ad hoc agent has the manipulation ability, which is illustrated in Fig. 6.10c. The proposed framework can adapt to ad hoc agents with different abilities, and the results demonstrate that the ad hoc agent with the manipulation ability can better help the original teams complete the task, obviously improving efficiency.

6.4.7 Qualitative Performance

We show the sample of the ad hoc agent collaborating with the LLM-based original team and compare the collaboration performance of *IRoT-LLM* ad hoc policy with *Naive-LLM* policy, which is demonstrated in Fig. 6.11. The icons of different colors mark positions of agents. Yellow lines with different shapes mark the paths between key steps of teammates, and white lines mark the paths of the ad hoc agent. A main timeline from 0 to 207-th step is illustrated at the top. The first row represents the task execution process of the original team in the top-down view. The second row shows the execution process of a new team with the *Naive-LLM* ad hoc agent joining at the 50-th step. The third row outlines the process of another new team with the *IRoT-LLM* ad hoc agent also joining at the 50-th step. The original team tidies up these three misplaced objects at the 207-th step. The new team with the *Naive-LLM* ad hoc agent finishes this task at the 154-th step, while the new team with the *IRoT-LLM* ad hoc agent finishes this task at the 136-th step. *IRoT-LLM* performs better than *Naive-LLM*, since *Naive-LLM* predicts the next sub-task without evaluation and reflection when replanning. *Naive-LLM* switches to *Explore* at the 90-th step when replanning after a teammate stops and then switches back to the sub-task *Replace*

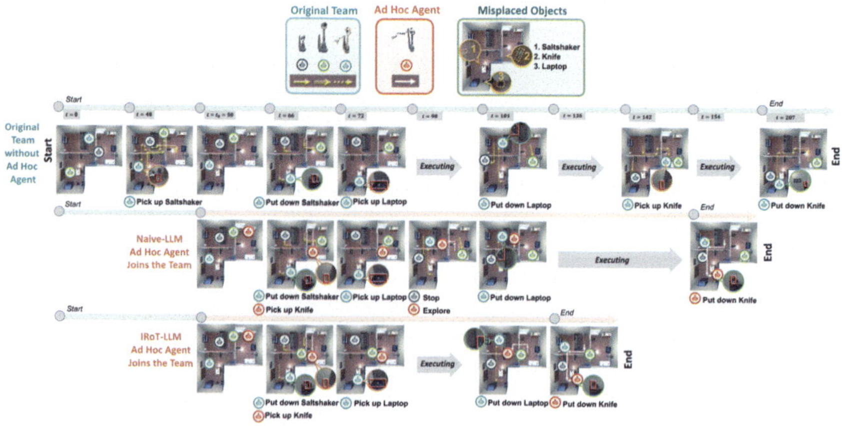

Fig. 6.11 The sample of the ad hoc agent collaborating with the LLM-based original team

after a few steps, while the *IRoT-LLM* one continues executing the sub-task *Replace* until it puts down the object, which demonstrates the effectiveness of the proposed IRoT method.

The original team contains three heterogeneous agents and completes the task with three misplaced objects at the 207-th step. The ad hoc agent joins the team at the 50-th step, and the actions of teammates are the same as the original team before the ad hoc agent joins the team. The new team with the ad hoc agent completes the task earlier than the original team. The new team with *IRoT-LLM* ad hoc agent finishes the same task earlier than the *Naive-LLM* policy, since *Naive-LLM* generates the executable sub-task without evaluation and reflection when replanning, which demonstrates the effectiveness of the IRoT method.

We also show the successful example of the proposed ad hoc teamwork based on IRoT in the simulation scene with both the first-person view and the top-down view in Fig. 6.12. Yellow lines with different shapes are used to mark the paths between key steps of teammates, and white lines are used to mark the paths of the ad hoc agent. The ad hoc agent joins the original at the 50-th step and helps the original team to complete the task. The ad hoc agent with the manipulation ability joins the original at the 50-th step. It can help the original team and replace the misplaced *Knife* to *CounterTop* in the kitchen. The joining of this ad hoc agent improves the efficiency of the original team.

6.4.8 Ablation Experiments

We conduct ablation experiments to evaluate the effectiveness of the proposed hierarchical dynamic planner and IRoT-based sub-task planner. The comparison

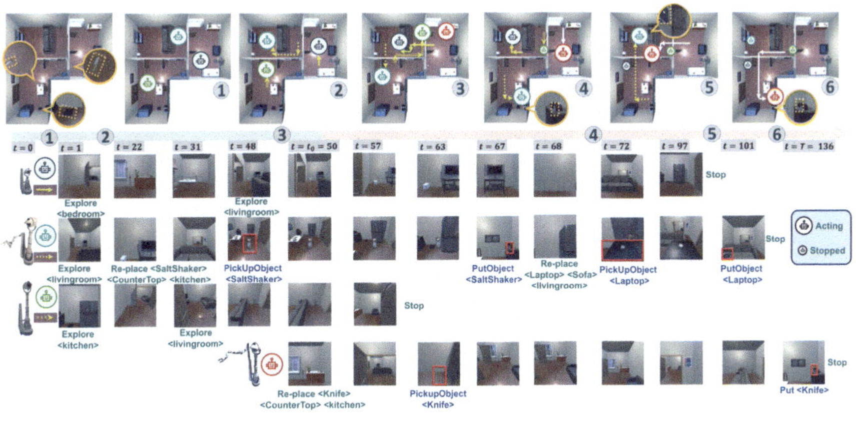

Fig. 6.12 The successful examples of the *IRoT-LLM* ad hoc agent. The icons of different colors mark positions of agents

performance of joining the team at different times and collaborating with LLM-based teammates is shown in Table 6.4. The method *w/o STP.* removes the dynamic sub-task planner and directly generates the sub-task with the LLM. The performance has decreased obviously compared with our framework, which demonstrates that hierarchical planning is effective in decision-making and can avoid invalid planning. We also discuss the effectiveness of different components of the proposed IRoT methods in the sub-task planning. *IRoT w/o Ev. & Rf.* does not utilize the evaluation and ranking as well as the interactive reflection steps, and it directly generates n_{irot} candidate sub-task and then chooses to execute them in the order of generation. *IRoT w/o Ev.* and *IRoT w/o Rf.* remove the evaluation and reflection steps, respectively. The comparison results show that the three steps of the proposed IRoT method are all useful, and each step can provide effective information for sub-task planning. *IRoT(GPT-3.5)* utilizes the *gpt-3.5-turbo-1106* model to complete the planning, and the performance demonstrates that the reasonable planning method is crucial for the LLM to fulfill its remarkable potential in planning and decision-making. The results of the ablation experiments show the effectiveness and efficiency of the proposed hierarchical planning with the IRoT planner.

Table 6.4 Results in ablation experiments of ad hoc teamwork

t_0	Ad hoc policy	$\%SR(\uparrow)$	$\%PS(\uparrow)$	$\#TS(\downarrow)$	$\#AS(\downarrow)$	$\%SE(\uparrow)$
0	No ad hoc	21.1 (-)	27.2 (-)	453.2 (-)	1813.3 (-)	–
	w/o STP.	21.9 (3.8)	27.8 (2.2)	441.3 (2.6)	2183.2 (-20.4)	0.6
	IRoT w/o Ev. & Rf.	22.5 (6.6)	28.3 (4.0)	423.2 (6.6)	2102.3 (-15.9)	1.5
	IRoT w/o Ev.	23.9 (13.3)	31.8 (16.9)	417.3 (7.9)	2013.9 (-11.1)	1.9
	IRoT w/o Rf.	23.6 (11.8)	31.3 (15.1)	418.8 (7.6)	2020.1 (-11.4)	1.8
	IRoT (GPT-3.5)	25.0 (18.5)	33.2 (22.1)	403.9 (10.9)	1956.3 (-7.9)	2.7
	IRoT-LLM	**26.3 (24.6)**	**35.1 (29.0)**	**399.3 (11.9)**	**1918.1 (-5.8)**	**3.2**
50	No ad hoc	21.1 (-)	27.2 (-)	435.2 (-)	1813.3 (-)	–
	w/o STP.	21.6 (2.4)	27.7 (1.8)	446.1 (1.6)	2132.8 (-17.6)	0.4
	IRoT w/o Ev. & Rf.	22.3 (5.7)	28.0 (3.0)	441.3 (2.6)	2050.6 (-13.1)	0.6
	IRoT w/o Ev.	23.5 (11.4)	31.3 (15.1)	425.2 (6.2)	1965.7 (-8.4)	1.4
	IRoT w/o Rf.	23.4 (10.9)	31.0 (14.0)	427.1 (5.8)	1970.7 (-8.7)	1.3
	IRoT (GPT-3.5)	24.8 (17.5)	33.0 (21.3)	408.2 (9.9)	1903.1 (-5.0)	2.5
	IRoT-LLM	**25.6 (21.3)**	**34.2 (25.7)**	**403.9 (10.9)**	**1859.2 (-2.5)**	**2.6**
100	No ad hoc	21.1 (-)	27.2 (-)	435.2 (-)	1813.3 (-)	–
	w/o STP.	21.5 (1.9)	27.5 (1.1)	450.8 (0.5)	2103.5 (-16.0)	0.2
	IRoT w/o Ev. & Rf.	22.0 (4.3)	27.7 (1.8)	446.6 (1.5)	2006.1 (-10.6)	0.3
	IRoT w/o Ev.	22.9 (8.5)	20.8 (13.2)	429.8 (5.2)	1911.3 (-5.4)	1.2
	IRoT w/o Rf.	22.7 (7.6)	20.6 (12.5)	431.2 (4.9)	1920.9 (-5.9)	1.1
	IRoT (GPT-3.5)	23.8 (12.8)	31.8 (16.9)	412.3 (9.0)	1856.2 (-2.4)	2.2
	IRoT-LLM	**25.0 (18.5)**	**33.1 (21.7)**	**407.6 (10.0)**	**1803.1 (0.6)**	**2.6**

6.5 Summary

In this chapter, we utilize the large language model (LLM) to extend the learning strategy of embodied multi-agent system to the training-free strategy. We focus on the heterogeneous ad hoc teamwork collaboration, which mainly considers generating the reasonable policy of an ad hoc agent to collaborate with unknown teammates without prior coordination. We discuss a more general and challenging ad hoc teamwork setting where the ad hoc robot can seamlessly join any team at any time from any location. We leverage the strong reasoning and generalization abilities of the LLM and propose the Interactive Reflection of Thought method to dynamically plan the sub-task and subskill for the ad hoc robot to execute. We also build a testing benchmark to evaluate the proposed ad hoc teamwork framework in the heterogeneous multi-agent tidying-up task. Extensive comparison and ablation experiments are conducted in the benchmark to demonstrate the effectiveness of the proposed framework.

References

1. Albrecht SV, Crandall JW, Ramamoorthy S: E-hba: Using action policies for expert advice and agent typification. In: Workshops at the twenty-ninth AAAI conference on artificial intelligence (2015)
2. Albrecht SV, Crandall JW, Ramamoorthy S (2016) Belief and truth in hypothesised behaviours. Artif Intell 235:63–94
3. Barrett S, Rosenfeld A, Kraus S, Stone P (2017) Making friends on the fly: cooperating with new teammates. Artif Intell 242:132–171
4. Chen S, Andrejczuk E, Cao Z, Zhang J (2020) AATEAM: achieving the ad hoc teamwork by employing the attention mechanism. In: AAAI, vol 34, pp 7095–7102
5. Chen Y, Cui W, Chen Y, Tan M, Zhang X, Zhao D, Wang H (2023) RoboGPT: an intelligent agent of making embodied long-term decisions for daily instruction tasks. arXiv preprint arXiv:2311.15649
6. Deitke M, VanderBilt E, Herrasti A, Weihs L, Salvador J, Ehsani K, Han W, Kolve E, Farhadi A, Kembhavi A, Mottaghi R (2022) ProcTHOR: large-scale embodied AI using procedural generation. In: NeurIPS
7. Dodampegama H, Sridharan M (2023) Back to the future: toward a hybrid architecture for ad hoc teamwork. In: AAAI, vol 37, pp 3–10
8. Howard A, Parker LE, Sukhatme GS (2006) Experiments with a large heterogeneous mobile robot team: exploration, mapping, deployment and detection. Int J Robot Res 25(5-6):431–447
9. Huang W, Abbeel P, Pathak D, Mordatch I (2022) Language models as zero-shot planners: extracting actionable knowledge for embodied agents. In: International conference on machine learning, pp 9118–9147. PMLR, New York
10. Huang W, Xia F, Xiao T, Chan H, Liang J, Florence P, Zeng A, Tompson J, Mordatch I, Chebotar Y, et al (2022) Inner monologue: embodied reasoning through planning with language models. arXiv preprint arXiv:2207.05608
11. Jain U, Weihs L, Kolve E, Farhadi A, Lazebnik S, Kembhavi A, Schwing A: A cordial sync: going beyond marginal policies for multi-agent embodied tasks. In: ECCV, pp 471–490. Springer, Berlin (2020)

12. Jain U, Weihs L, Kolve E, Rastegari M, Lazebnik S, Farhadi A, Schwing AG, Kembhavi A: Two body problem: collaborative visual task completion. In: Proceedings of the IEEE/CVF conference on computer vision and pattern recognition, pp 6689–6699 (2019)
13. Ju C, Son HI (2019) Modeling and control of heterogeneous agricultural field robots based on Ramadge–Wonham theory. IEEE Rob Autom Lett 5(1):48–55
14. Liu X, Guo D, Liu H, Sun F (2022) Multi-agent embodied visual semantic navigation with scene prior knowledge. IEEE Rob Autom Lett 7(2):3154–3161
15. Liu X, Li X, Guo D, Tan S, Liu H, Sun F (2022) Embodied multi-agent task planning from ambiguous instruction. In: Proceedings of robotics: science and systems, New York City, NY, USA, pp 1–14
16. Liu Z, Bahety A, Song S (2023) Reflect: summarizing robot experiences for failure explanation and correction. arXiv preprint arXiv:2306.15724
17. Miao R, Jia Q, Sun F: Long-term robot manipulation task planning with scene graph and semantic knowledge. Rob Intell Autom 43(1):12–22 (2023)
18. Mirsky R, Carlucho I, Rahman A, Fosong E, Macke W, Sridharan M, Stone P, Albrecht SV (2022) A survey of ad hoc teamwork: definitions, methods, and open problems. In: European conference on multiagent systems
19. Prasad A, Koller A, Hartmann M, Clark P, Sabharwal A, Bansal M, Khot T (2023) Adapt: As-needed decomposition and planning with language models. arXiv preprint arXiv:2311.05772
20. Qiao Y, Qi Y, Yu Z, Liu J, Wu Q (2023) March in chat: interactive prompting for remote embodied referring expression. In: Proceedings of the IEEE/CVF international conference on computer vision, pp 15758–15767
21. Rahman MA, Hopner N, Christianos F, Albrecht SV (2021) Towards open ad hoc teamwork using graph-based policy learning. In: International conference on machine learning, pp 8776–8786. PMLR, New York
22. Sarch G, Wu Y, Tarr MJ, Fragkiadaki K (2023) Open-ended instructable embodied agents with memory-augmented large language models. arXiv preprint arXiv:2310.15127
23. Sarıçiçek İ, Yazıcı A, Aslan Ö (2023) A novel conflict free routing with multi pickup delivery tasks for autonomous vehicles. Rob Intell Autom 43(2):191–206
24. Schuster MJ, Müller MG, Brunner SG, Lehner H, Lehner P, Sakagami R, Dömel A, Meyer L, Vodermayer B, Giubilato R, et al (2020) The arches space-analogue demonstration mission: towards heterogeneous teams of autonomous robots for collaborative scientific sampling in planetary exploration. IEEE Rob Autom Lett 5(4):5315–5322
25. Sharma V, Goyal P, Lin K, Thattai G, Gao Q, Sukhatme GS (2022) CH-MARL: a multimodal benchmark for cooperative, heterogeneous multi-agent reinforcement learning. arXiv preprint arXiv:2208.13626
26. Shridhar M, Thomason J, Gordon D, Bisk Y, Han W, Mottaghi R, Zettlemoyer L, Fox D (2020) ALFRED: a benchmark for interpreting grounded instructions for everyday tasks. In: Proceedings of the IEEE/CVF conference on computer vision and pattern recognition, pp 10740–10749
27. Stone P, Kaminka G, Kraus S, Rosenschein J (2010) Ad hoc autonomous agent teams: collaboration without pre-coordination. In: AAAI, vol 24, pp 1504–1509
28. Sudarshan SK, Montano V, Nguyen A, McClimans M, Chang L, Stewart, RR, Becker AT (2017) A heterogeneous robotics team for large-scale seismic sensing. IEEE Rob Autom Lett 2(3):1328–1335
29. Tan S, Xiang W, Liu H, Guo D, Sun F (2020) Multi-agent embodied question answering in interactive environments. In: ECCV, pp 663–678. Springer, Berlin
30. Wang H, Wang W, Zhu X, Dai J, Wang L (2021) Collaborative visual navigation. arXiv preprint arXiv:2107.01151
31. Wang M, Cong M, Du Y, Liu D, Tian X (2023) Multi-robot raster map fusion without initial relative position. Rob Intell Autom 43(5):498–508
32. Wei J, Wang X, Schuurmans D, Bosma M, Xia F, Chi E, Le QV, Zhou D, et al (2022) Chain-of-thought prompting elicits reasoning in large language models. Adv Neural Inf Proces Syst 35:24824–24837

33. Weihs L, Deitke M, Kembhavi A, Mottaghi R (2021) Visual room rearrangement. In: Proceedings of the IEEE/CVF conference on computer vision and pattern recognition, pp 5922–5931
34. Yang W, Wang X, Farhadi A, Gupta A, Mottaghi R (2018) Visual semantic navigation using scene priors. arXiv preprint arXiv:1810.06543
35. Yao S, Zhao J, Yu D, Du N, Shafran I, Narasimhan K, Cao Y (2022) React: synergizing reasoning and acting in language models. arXiv preprint arXiv:2210.03629
36. Yu C, Yang X, Gao J, Yang H, Wang Y, Wu Y (2022) Learning efficient multi-agent cooperative visual exploration. In: ECCV, pp 497–515. Springer, Berlin
37. Zintgraf L, Devlin S, Ciosek K, Whiteson S, Hofmann K (2021) Deep interactive bayesian reinforcement learning via meta-learning. arXiv preprint arXiv:2101.03864

Part III
Applications

This part of the book focuses on the application examples. It comprises two chapters. Chapter 7 demonstrates a comprehensive human-robot simulation platform, and Chap. 8 presents two typical practical deployment examples.

Chapter 7
Simulation Platform for Embodied Collaboration Between Human and Robots

Abstract Human-robot collaboration (HRC) in shared workspaces has become increasingly common in real-world applications and has drawn considerable research attention. However, most existing studies on human-in-the-loop (HITL) collaboration with robots are conducted either in simplified game environments or on physical platforms, which limits their realism and scalability. To address this, we propose an embodied framework called HumanTHOR, which allows humans to interact with robots in a simulation environment using VR devices, facilitating HITL collaboration in a shared workspace. To validate the system, we develop a benchmark of everyday tasks and conduct a preliminary user study using two baseline algorithms. The results demonstrate that the robot can effectively assist humans in collaboration, highlighting the significance of HRC. A comparison of different baseline levels confirms that our system can assess robot capabilities and serve as a benchmark for evaluating various robot algorithms. Additionally, the experimental results suggest that there is still substantial potential in this area, and our system provides a solid foundation for future HRC research in shared workspaces.

7.1 Introduction

With advancements in multimodal perception, reasoning, planning, and control, robots have become increasingly capable and are gradually being deployed across a wide range of application scenarios. Human-robot collaboration (HRC) has gained significant attention and become a focal point in various real-world domains, including industry [22, 45], surgery [19, 29], rescue [31], healthcare [3], agriculture [44], and home services [23].

A common and realistic collaboration scenario is loosely coupled teamwork in a shared workspace, such as household chores, where individuals work independently toward a shared goal and may communicate with the robot to exchange information. Studying HRC in such a scenario requires two key features of the simulation system: (1) The task is human in the loop (HITL), necessitating real-time interaction with humans. (2) The human and robot operate in a shared workspace, where they have a

H. Liu et al., *Embodied Multi-Agent Systems*, Machine Learning: Foundations, Methodologies, and Applications, https://doi.org/10.1007/978-981-96-5871-8_7

similar level of observation and capabilities, rather than following an instructor-follower paradigm as seen in previous works [16, 33]. For instance, in the task of placing an apple in the fridge, the human first searches for the apple and then searches for the fridge to store it. In a human-robot collaborative scenario, the human and robot can simultaneously search for the apple and fridge in the shared workspace. Once the robot locates the fridge, it can inform the human of its position, allowing the human to directly transport the apple to the fridge, thereby improving work efficiency.

To advance research in this field, it is essential to establish a robust system and corresponding benchmarks. However, existing systems may fall short in supporting HITL collaboration with robots in a shared workspace due to several limitations. One common approach is to study HRC using physical systems, such as collaborative furniture assembly [30], pick-and-place tasks [13], and simple object picking [36]. While these physical systems are practically effective, they often suffer from high economic costs, limited reproducibility, and relatively small scales. Alternatively, research on cooperation in 2D games, such as Overcooked [4] and the cooperative table-carrying game [32], may be overly simplified and not generalizable to physical robots, despite their greater reproducibility.

Fortunately, recent advancements have introduced several embodied simulators, such as the AI2THOR series [9, 12, 21], which serve as excellent platforms for studying robot algorithms. These simulators are capable of scaling up and closely approximating real-world scenarios. However, despite the recognition of the importance of human factors and the expansion of benchmarks to incorporate various modalities of human-collected data, as seen in studies such as [16, 17, 49], there have been limited efforts to study HITL collaboration within these embodied simulators.

To facilitate the study of HITL collaboration with robots in a shared workspace, we develop an embodied human-robot collaboration environment based on the AI2THOR framework, named HumanTHOR. As illustrated in Fig. 7.1, humans can

Fig. 7.1 An overview of the HumanTHOR system

immersively control their avatars using VR devices to collaborate with the robot in the environment. Unlike previous works that use VR to collect human demonstrations [14, 24, 25], our environment is specifically designed for HRC tasks, offering various functional supports such as enabling synchronous collaboration between humans and robots, as well as providing image-text communication interfaces.

To demonstrate our system, we implement a preliminary HRC benchmark for everyday tasks, including two representative tasks: *object goal navigation* and *mobile manipulation*. To create realistic and diverse episodes, initial object arrangements are sampled from a set of scene priors. We validate our system through a user study, using a rule-based robot and an oracle robot as baselines. Experimental results show that robots can significantly assist humans in completing everyday tasks, and our system proves to be an effective testbed for HRC research.

7.2 Related Work

7.2.1 Human-Robot Collaboration

The quest to create collaborative and user-friendly robots has garnered significant attention in the research community. To enhance user experience, Virtual Reality (VR) has become a popular user interface in human-robot interaction studies [11]. Beyond serving as a display medium [34], VR controllers enable humans to teleoperate robots or their avatars [8, 20, 42]. These VR-based controls are also widely used for collecting human demonstrations to aid robot learning [15, 24].

Since VR offers virtual and digital experiences, these techniques can also be integrated with real-world systems to form Augmented Reality (AR) or Mixed Reality (MR), which significantly enhance collaborations in the physical world. MR can act as a communication medium to convey the intentions of robots [30, 36], while AR can facilitate the study of human-robot collaboration in shared workspaces [35].

To enable immersive environment exploration and operations, and to closely approximate real-world human-robot collaboration, we utilize VR as the user interface in our HumanTHOR system.

7.2.2 Simulators and Benchmarks for Embodied Intelligence

Along with the rising interest in embodied intelligence, a considerable number of simulation environments emerge. Environments such as AI2THOR series [9, 12, 21], iGibson [24, 39], Habitat [38, 43], Matterport3D [5], and RFUniverse [15] have present vivid simulation with various physics systems and make it possible for building benchmark for tasks such as navigation or manipulation.

Beyond the platforms, various benchmarks for embodied tasks have been proposed, such as question answering [7], visual language navigation [1], scene graph generation [26], and performing composite everyday tasks [40]. After achieving adorable success in these tasks, researchers have turned their gaze to multi-agent settings [27, 28], considering human factors [41, 50], and collaboration with humans. Beyond a single paragraph of instructions, GesTHOR [49] augments the embodied navigation with human gesture indications. HandMeThat [46] takes one more step toward HRC by playing a record of human actions and instructions and then trains the robots to follow such instructions. Recent benchmarks consider communication between humans and robots. DialFRED [17] and TEACh [33] extend the instructions into dialogue form and encourage the robots to actively raise questions. Alexa Arena [16] builds an interactive environment where the instructor and executor can interactively communicate with natural language. While the benchmark is built on offline data, the environment may also support HITL collaborations.

To comprehensively support our intended HRC study, the system must include several key features: (1) **Vivid 3D simulation**, which aligns the simulation with real-world robot applications and enhances the human experience; (2) **HITL interaction**, enabling experiments with real humans rather than proxy agents or offline data; (3) **Shared workspace collaboration**, which is central to the domain we aim to study; and (4) **Instant communication**, essential for human-robot interaction and simulating real-world collaboration. For loosely coupled collaboration, where humans and robots may not always be within each other's line of sight, the system should support multimedia-style communication, incorporating both images and text. Additionally, the system should feature a VR interface, providing an immersive experience for humans and better aligning human activities with real-world scenarios.

However, existing simulation environments fall short of meeting all the requirements outlined above. To address this gap, we have developed the HumanTHOR system, which integrates all of these features. A detailed comparison between the proposed HumanTHOR and existing simulation environments is provided in Table 7.1.

7.3 System Framework

7.3.1 System Overview

Aiming to support human-in-the-loop collaborations in a shared workspace with flexible and multimodal communication, we design and implement the HumanTHOR system based on the AI2THOR platform. As the hierarchical architecture illustrated in Fig. 7.2, our system consists of three components:

- The **THOR simulator** implemented with the Unity framework based on the AI2THOR infrastructure, responsible for physical simulation, scene rendering,

Table 7.1 Comparison between HumanTHOR and related HRC benchmarks across various aspects

Benchmark	Simulation domain	VR interface	HITL collaboration	Shared workspace	Communication	
					Image	Text
Overcooked [4]	2D	✗	✓	✓	✗	✗
It takes two [32]	2D	✗	✓	✓	✗	✗
HandOverSim [6]	3D	✗	✗	✓	✗	✗
VR Kitchen [18]	3D	✓	✗	✗	✗	✗
DialFred [17]	3D	✗	✗	✗	✗	✓
TeaCh [33]	3D	✗	✗	✗	✗	✓
HandMeThat [46]	3D	✗	✗	✗	✗	✓
GesTHOR [49]	3D	✓	✗	✗	✗	✗
Alexa Arena [16]	3D	✗	✓	✗	✗	✓
HumanTHOR	3D	✓	✓	✓	✓	✓

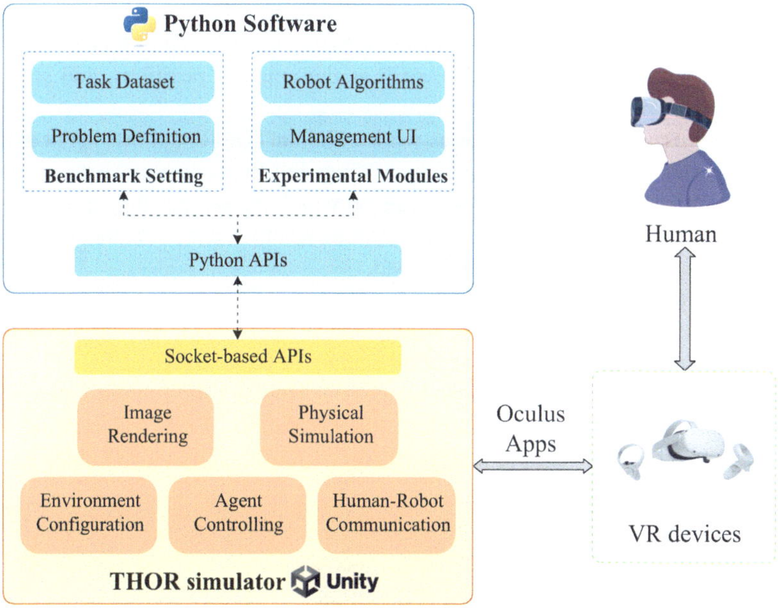

Fig. 7.2 The architecture of our HumanTHOR system

and solving the interaction between agents and environments. The simulator exposes a set of socket-based APIs for controlling.

- The **Python software** that supports further HRC study, including extensible benchmarks which will be introduced in Sect. 7.4, a code framework for running robot algorithms and a web user interface for controlling and monitoring the HRC experiments.

- **VR devices**. We use Meta Quest 2 in our system, including a headset and a pair of controllers. VR devices are connected to our simulator through the Oculus platform.

By integrating VR devices into our system, humans and robots can concurrently work within the environment, thereby achieving HITL and human-robot collaboration in a shared workspace.

7.3.2 Supported APIs

To meet the diverse requirements of HRC benchmarks, our system offers a wide range of APIs within the HumanTHOR environment, which can be broadly categorized into the following four groups:

- **Environment configuration**: We provide a series of APIs for setting up the environment, including *selecting scenes*, *initialization*, and *object settings*. To support the development of benchmarks to set up customized tasks, environmental information is also obtainable, such as the amount, positions, and states of objects.
- **Agent control**: Similar to previous works, HumanTHOR supports agent actions such as *move*, *rotate*, *teleport*, *pick*, and *place*. For the convenience of customizing benchmarks, we use a flexible format in which the *move* action and the *rotate* action take a vector that indicates the variation as input, and leave the designation of atomic actions to the benchmark level.
- **Perception and monitoring**: Following the conventions, our environment provides an ego-centric RGB observation and a depth map. We also provide a top-view image and the positions of all agents for monitoring.
- **Communication**: A major contribution of our HumanTHOR environment is the multimodal communication module for collaboration. We provide communication APIs including sending image-text messages, responding to messages, and querying the response status.

7.3.3 User Interface and VR Supports

To deliver an immersive experience for human participants, we utilize VR as the user interface. With the VR headset, humans can view the environment from a first-person perspective and control their avatars using the buttons and joysticks on the controllers, allowing them to navigate the environment and interact with objects. Additionally, we have developed several interfaces for human-robot communication, including an image-text message box and a location map, both of which are visible in the human's field of view. These operations are demonstrated in Fig. 7.3. In our collaborative task benchmark, the message appears in a dialogue box within

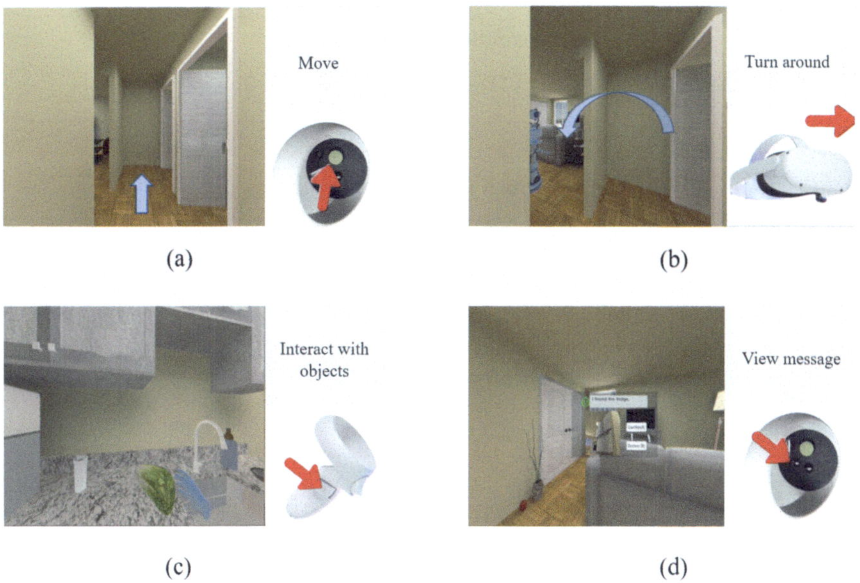

Fig. 7.3 Interacting with the environment using VR devices: (**a**) The human avatar can be moved by operating the joystick on the VR controller. (**b**) The head-mounted display's sensors allow humans to easily rotate their viewpoint by physically turning around. (**c**) When close to a movable object, humans can pick it up by pressing the side button on the controller, or open a receptacle, such as a fridge. (**d**) Upon receiving a message from the robot, humans can quickly respond using the A/B buttons on the controller

the human's view, where button A is for confirmation and button B is for decline. Once confirmed, a map showing the relative positions of the robot and human is displayed, and it can be hidden or redisplayed using the same button operation.

7.4 Benchmarking HRC with Everyday Tasks

The HumanTHOR platform is eligible for studying various HRC tasks at various levels, being available for both evaluating robot algorithms and human-robot collaboration studies. We model the benchmark structure in a hierarchical way as illustrated in Fig. 7.4, where we start by supporting elemental tasks, based on which we can perform practical everyday tasks, such as visual language navigation [40], room rearrangement [2, 47], and tidying-up [37]. Ultimately, HumanTHOR supports studying high-level scientific problems such as Theory-of-Mind and human trust modeling.

To preliminarily validate the effectiveness of our system, we build benchmarks for the elementary **object goal navigation** task and a compound HRC task, **mobile manipulation**, as defined in the following:

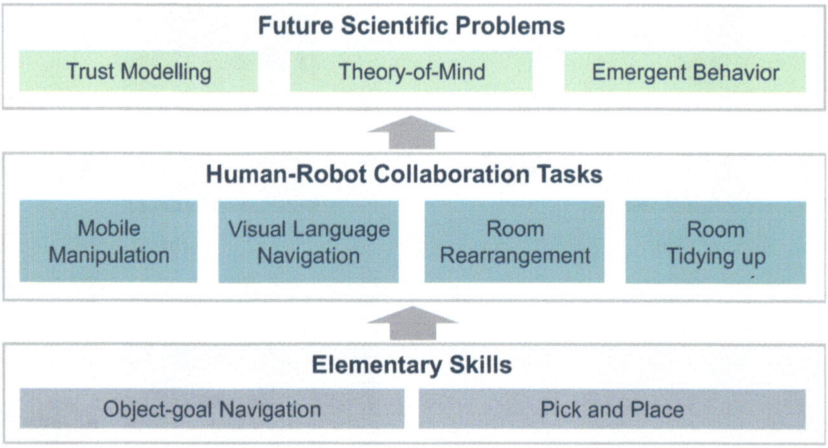

Fig. 7.4 The hierarchical benchmarks supported by the HumanTHOR platform

- **Object goal navigation**, a representative and fundamental task in the embodied intelligence domain. Such tasks target a common situation in which a human is searching for something in the house. The task succeeds when the human successfully finds the target object.
- **Mobile manipulation**. In this task, participants need to pick up the target object and place it at the destination place. This is a fundamental type of everyday task, which is also adopted by ALFRED [40], being more complex than navigation tasks since it involves multiple objects and interactions with the environment.

Two examples of tasks are presented in Fig. 7.5. Similar to previous works [16, 40], our tasks are defined with three essential components: **initial state** that determines the initial arrangement of objects, **goal definition** that formalizes the criterion of success, and **auxiliary information** such as natural language description of the tasks.

7.4.1 Object-Centric Interactive Communication

We implement an object-centric communication flow with corresponding user interfaces to facilitate human-robot collaboration in everyday tasks. When the robot deems it necessary to communicate object-related information, it can send an image-text message as shown in Fig. 7.6a. This message includes the robot's observation along with a custom text message, as illustrated in Fig. 7.6b. If the human is interested in the message, they can confirm it by pressing a button. Following this, a map showing the relative positions of the human and the robot will be displayed,

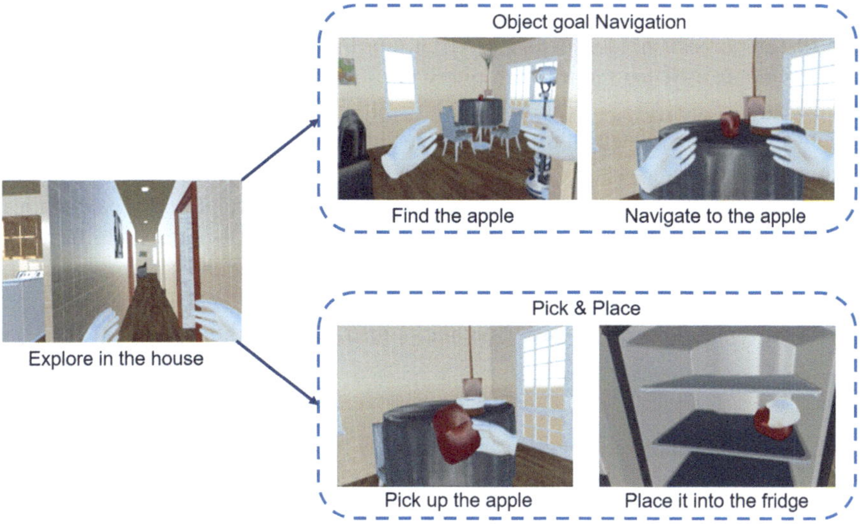

Fig. 7.5 The general processes of navigation tasks and manipulation tasks

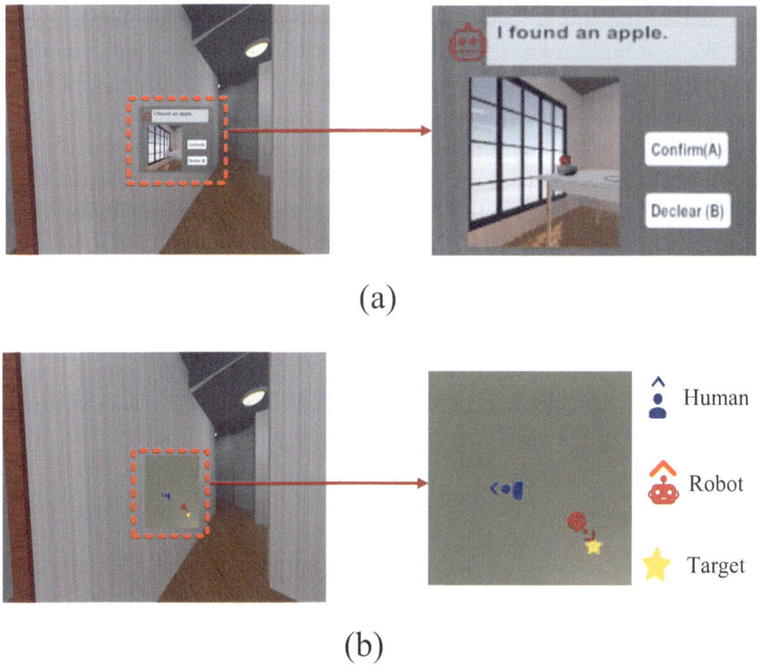

Fig. 7.6 An illustration of the object-centric communication for human-robot collaboration. (**a**) The image-text message sent by the robot. (**b**) The map with relative locations after confirmation

assisting the human in navigating to the suggested location. Both user interfaces can be temporarily hidden or displayed at any time by the human. Additionally, we provide an API for the robot to query whether its suggestion was accepted or declined.

It is important to note that robots have the autonomy to choose when and what to communicate in their messages, with the ability to override previous messages with new ones. This flexibility, when combined with the hierarchical levels of capabilities in heterogeneous teamwork, leaves substantial room for further research. In more complex tasks, robots may need to carefully assess the human's mental state to determine the optimal timing for sending messages. The challenge lies in conveying crucial information without overwhelming or distracting the human too much.

7.4.2 Generating Everyday Tasks with Scene Priors

Our benchmark incorporates 10 scenes and 14 objects, including eight receptacle objects and six target objects. We obtain 19 mobile manipulation task templates by artificial annotations, which are in the form of *Picking a/an {target object} and place it in/on a/an {receptacle object}*.

We generate the initial state with the guide of scene priors like exhibited in Fig. 7.7. We first build a knowledge graph with 65 triplets by human annotations, and each triplet represents a possible prior relation $(target, relation, reference)$. We weight the triplets by combining the similarities between the textual features of objects adopted from the pretrained BERT model [10] and human priors.

We follow a three-step process to instantiate the task, as illustrated in Fig. 7.7. First, we select a task template that is relevant to the scene, which helps define the *goal*. Next, we build a scene graph by sampling scene priors from the knowledge graph. Each sampled relation corresponds to a designated area where the object should be placed, and the exact position of each object is determined accordingly. By following this procedure, we generate 565 navigation tasks and 1583 mobile manipulation tasks. The initial relations of the target object are sampled once for each task template in each scene, while the irrelevant objects are not constrained.

7.5 Experiments

To validate our proposed HumanTHOR system, we conduct a preliminary user study on a subset of our benchmark, focusing on the mobile manipulation tasks. These tasks are more complex and include the navigation task as a component. In this experiment, we aim to address two key questions: (1) the necessity of HRC: can the robot effectively assist humans in completing the tasks? (2) the effectiveness of

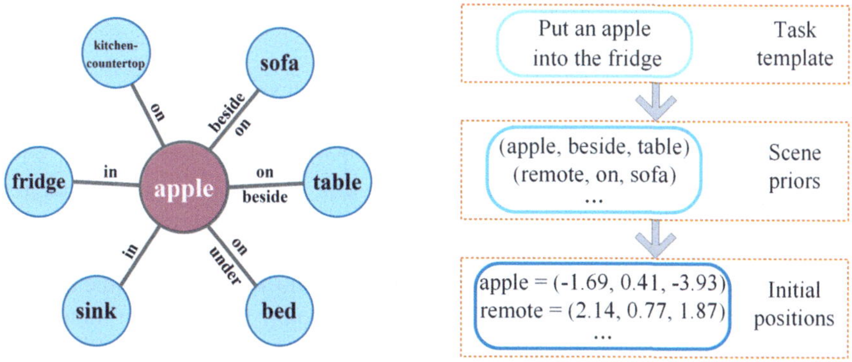

Fig. 7.7 An example of scene priors and the task generation pipeline

our system as an evaluation benchmark: can our experiment differentiate between robots of varying capabilities and evaluate their performance?

7.5.1 Baselines

We incorporate two simple robot baselines in the experiments. For both robots, the enabled actions include *move*, *rotate*, and *send message*. The atomic movement distance cannot exceed 0.5 meters, the angle of pitch is limited to $[-30°, 30°]$, and sending a message requires a custom text and an estimated position. The details of the baselines are as follows:

- Frontier agent. The Frontier agent is a rule-based agent that explores the environment using the frontier algorithm and sends messages when it detects the target. To simplify the task, we provide the Frontier agent with an ideal object detector, capable of detecting objects with ground truth accuracy within a 1.5-meter range.
- Oracle. The Oracle robot has knowledge of the positions of all objects and navigates to the target object along the shortest path. Despite its complete object knowledge, it still adheres to the communication framework and only sends messages when it is close enough to detect the target.

Both baselines are set to a default configuration, where the robot searches for the target object rather than the receptacle or dynamic task allocation. While these baselines may not be practical or optimal, they serve to verify and calibrate our system: The Oracle represents an approximation of the upper bound, and the Frontier serves as a rule-based baseline without the need for training or intelligent algorithms. Additionally, we set up a control group that has no robot to compare the human performance in the absence of robotic assistance.

7.5.2 Experimental Setup

We sample three mobile manipulation tasks from each scene for the experiments, resulting in a total of 30 tasks across 10 scenes. We invite 18 human participants, equally divided into three groups. Each group performs 10 random tasks, with one task in each scene. For consistency, the task sequence is randomized for each participant. Participants are given 90 seconds to complete each task. After completing each task, they are asked to rate their trust in the robot (if applicable) using a trust scale from 1 to 7, where 1 indicates no trust and 7 indicates the highest level of trust. While the participants do not know the specific robots, they are informed that they will be collaborating with the same robot throughout the experiment. After excluding any broken cases, we have 175 valid trials for analysis.

We assess the capability of robots through the success rate and execution time. The task success score s is defined as 1 if the target object is placed in the correct place, and 0 otherwise. The success rate (SR) is calculated by the ratio of success tasks. As a human-centric experiment, we use the time-weighted success rate (TWSR) instead of path-weighted success rate (PWSR). The TWSR score s_t is calculated by the following formula:

$$s_t = s * \frac{T^*}{\max(T^*, T)}, \tag{7.1}$$

where T denotes the time spent on the task, and T^* is the minimum required time that is estimated by the distance of the shortest path. Besides, we record the adoption rate for robots to measure their help. A robot is considered adopted if it sends a message that is confirmed by the human. Participants can modify their adoption orally when mistouching the button, or exclude it from statistics for boundary cases.

7.5.3 Results

7.5.3.1 Quantitative Analysis

We present the quantitative scores in Table 7.2. Compared to the control group where no robot assistant exists, both the frontier robot and the oracle bring improvement in all quantitative metrics, including the success rate and the time-weighted success rate. As an approximation of the upper limit, the oracle improves the success rate significantly to 91.4%, much more significant than the frontier agent which serves as a medium-level baseline. The oracle also has a much higher adoption rate, suggesting it is considered more helpful in collaboration. The significant discrepancies in quantitative metrics show that our system can successfully distinguish the capability of robots and thus be an eligible evaluation benchmark.

Table 7.2 Average success rate, time-weighted success rate, and adoption rate in all three settings

Robot setting	SR (%)	TWSR (%)	Adoption rate (%)
No robot	63.8	46.7	/
Frontier	72.9	54.3	56.0
Oracle	91.4	66.7	90.0

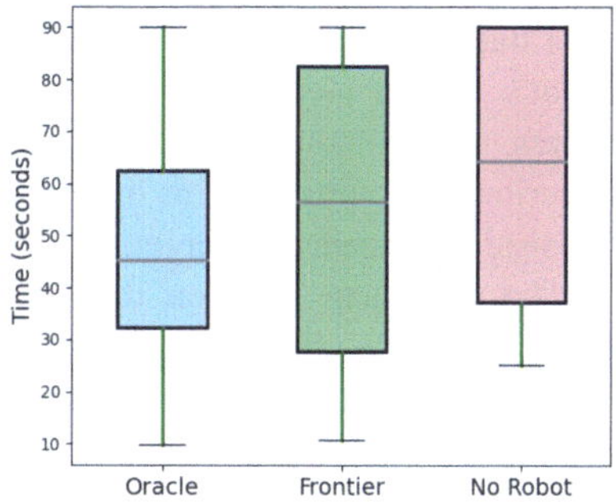

Fig. 7.8 Average time spent to complete the tasks

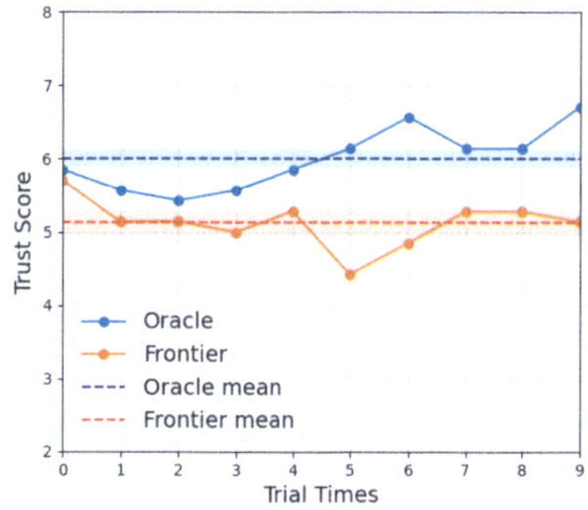

Fig. 7.9 The variation of average human trust scores over trials

Another representative indicator, the average time spent across tasks, is shown by the box plots in Fig. 7.8. Although the robot successfully improves the overall performance, we find the improvement mainly appears in the hard tasks. As illustrated, the lower bounds of time spent by the oracle and the frontier are almost the same, and the lower quartiles of all three settings do not have significant differences. This fits our insights that in household tasks, the robot plays its role when the task is at a specified level of difficulty. These are too easy tasks in which the targets are very close to the robot or the human cannot examine the capability of robots. As suggested by the results, our benchmark covers tasks of various levels of difficulty and thus can be used as a general evaluation platform.

7.5.3.2 Human Trust

We present the variation and average of human trust across trial times in Fig. 7.9, with shaded areas representing standard errors. Notably, the oracle consistently receives higher human trust than the frontier agent, confirming that humans recognize differences in robot capabilities and tend to trust more capable agents. An interesting observation is that since humans lack a top-down view and are unaware of the robot's trajectory, they often assess the robot's capability based on the time spent searching. This leads to fluctuations in trust scores. For instance, when a task is difficult and the target object is far from the robot, humans may lower their trust due to the robot's limited assistance. However, humans also account for the robot's overall performance, which is reflected in the gradually stabilizing trust scores. Additionally, we observe a rising trend in trust scores for the oracle as the experiment progresses. This likely occurs because humans recognize that the oracle consistently locates the target, while the frontier agent still has a chance of failure.

The clear gap between the frontier agent and the oracle demonstrates that our experiment effectively differentiates agent capabilities, making it a valuable benchmark for evaluating robot performance and studying human trust in robots.

7.5.4 Case Study

We present a case study in Fig. 7.10 to illustrate the general collaboration process. In this scenario, the human is tasked with finding the remote and placing it on the bed. The global maps display the layout of the house and the positions of task-related entities, with the human starting in the corridor and the robot located in the living room.

As shown by the trajectories, the oracle robot directly navigates to the target and is the first to report its position to the human. With this guidance, humans can quickly complete the task with almost no extra exploration. Without prior knowledge, the frontier agent must spend some time exploring and may even fail at times. However, it still successfully finds the target in many cases, saving humans

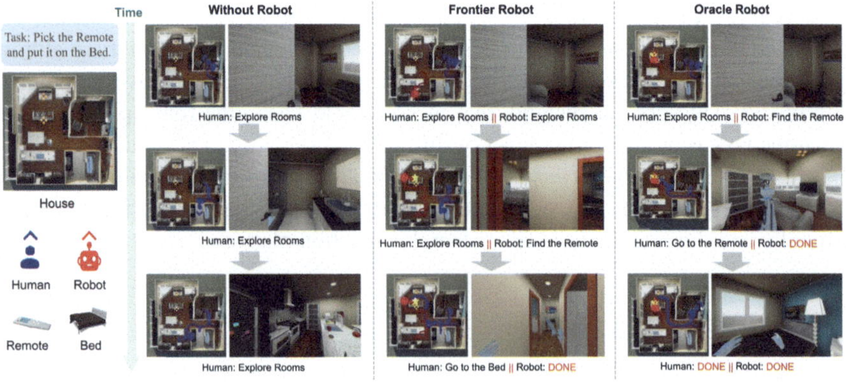

Fig. 7.10 A representative task for demonstrating the effect of robot assistants

some exploration time. In contrast, when there is no robot assistant, the human may need to explore the entire house and spend much more time. In one case, the human does not even find the target, while the oracle-guided participant succeeds, and the frontier-assisted participant retrieves the remote and returns to the bed. The comparison among these three trajectories demonstrates the varying levels of assistance provided by different robots in collaboration tasks.

Since the selected case represents a medium difficulty in a small house, all three trials succeed within nearly 75 seconds. However, robot assistance becomes even more crucial in more challenging tasks. For further examples and details, please refer to our accompanying video.

7.6 System Extensions

Based on the features introduced, our platform is highly extensible for customized complex tasks and studies. In this section, we discuss the extensibility of HumanTHOR, including additional support for multi-robot settings, more complex tasks, and the integration of various robot algorithms.

7.6.1 Mutli-Robot Setting

HumanTHOR supports the integration of multiple robots, of either the same type or heterogeneous, within a single scene. To demonstrate this capability, we present a multi-robot multi-target mobile manipulation task, as shown in Fig. 7.11. The complete task process is illustrated in the human's first perspective. Since the small robot cannot interact with the fridge, it reports to the human and lets the human check the fridge. In contrast, the tall robot with manipulation capability can directly take the apple back. In this task, the human is assisted by two robots with different capabilities: The taller robot possesses manipulation capabilities, while the smaller robot can only navigate. Both robots can communicate with the human and contribute to the task in distinct ways.

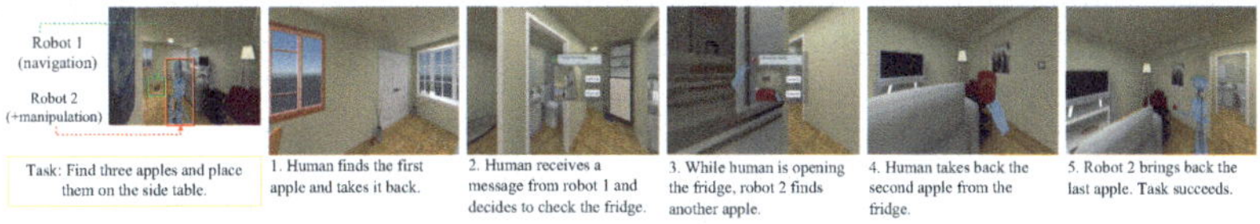

Fig. 7.11 A demonstration of HRC in a multi-robot multi-target mobile manipulation task, where the human-robot team is asked to collect and place three apples

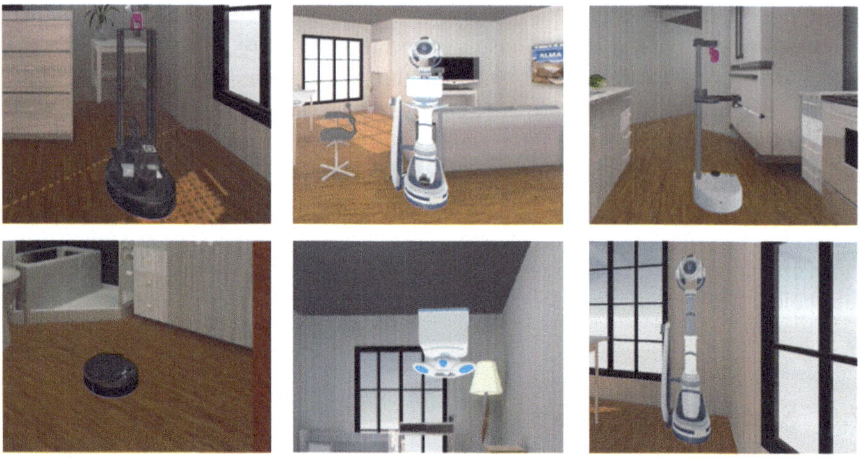

Fig. 7.12 The HumanTHOR system supports various types of robots and is extensible for further customization. Current robots are from the ProcTHOR dataset

The HumanTHOR system is also flexible and extensible to accommodate more types of robots. These robots can differ in appearance, size, performance, and capabilities. A few examples of currently available robots are shown in Fig. 7.12.

7.6.2 More Complex Tasks

As introduced in Sect. 7.3.2, the HumanTHOR simulator exposes APIs for configuring scenes and objects on the fly, enabling multistage tasks such as *room rearrangement* [2]. In this task, the human and robots can first tour the clean environment and record the object arrangement, as shown in Fig. 7.13a. Once they are ready, some object arrangement will be shuffled as in Fig. 7.13b, requiring rearrangement.

Similarly, our system supports *tidying up* [37], another rearrangement task where no reference environment is provided and the agents need to rearrange objects according to commonsense, making the communication between the human and robots more important. For *room rearrangement* and *tidying-up* tasks, we provide more details and corresponding demo videos with an oracle agent on our website.

7.6.3 More Robot Algorithms

As introduced in Sect. 7.3.2, the HumanTHOR simulator provides ego-centric RGB observations and depth maps, making it well-suited for vision-based navigation

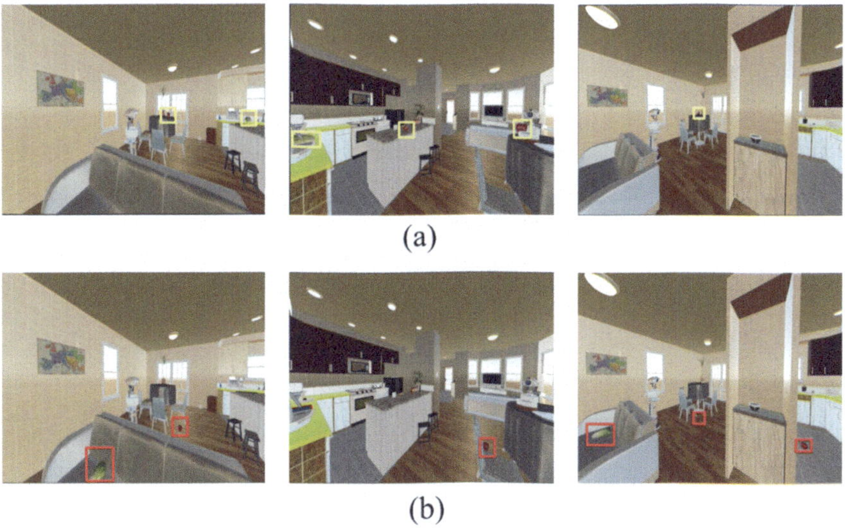

Fig. 7.13 Comparison between the original scene and the shuffled one within three angles of view in the *room rearrangement* task. (**a**) Original object arrangement. (**b**) Shuffled object arrangement

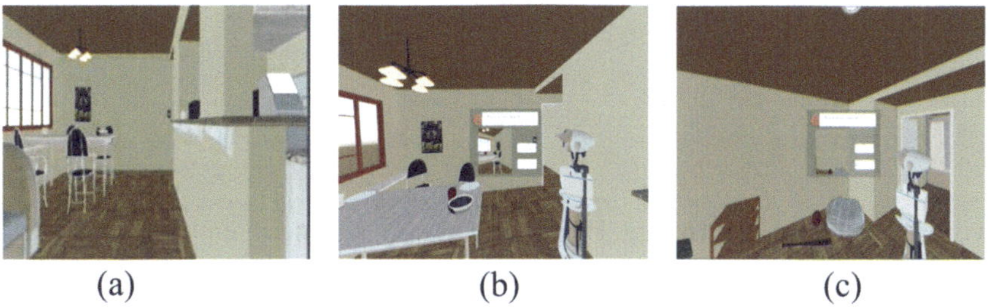

Fig. 7.14 An example of employing the SAVN model to search for an apple in our HumanTHOR system. (**a**) Robot observation. (**b**) Success case. (**c**) Failure case

algorithms. To showcase this capability, we implement a representative learning-based visual navigation model, SAVN [48]. As shown in Fig. 7.14, existing visual navigation models can be effectively applied within the HumanTHOR system. The platform has already been demonstrated in the RSS2024 conference in the Netherlands, as shown in Fig. 7.15. However, searching for objects across an entire house remains a challenging task, indicating that there is still room for further development and improvement.

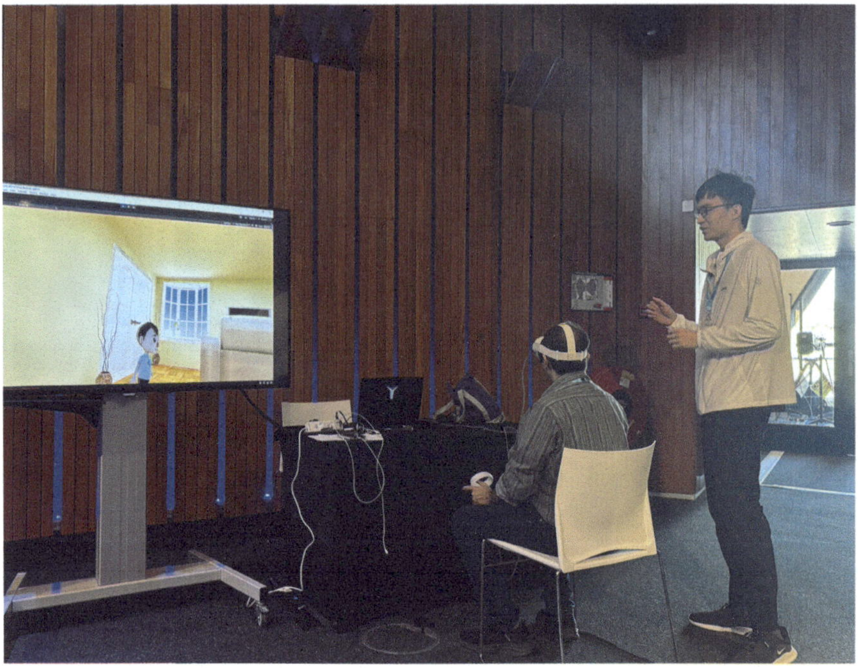

Fig. 7.15 Demonstration at RSS2024

7.7 Summary

In this chapter, we introduce HumanTHOR, an advanced embodied simulator equipped with an everyday task benchmark for studying human-robot collaboration in a shared workspace. Unlike existing environments and benchmarks, HumanTHOR offers not only a realistic simulation and a human-in-the-loop collaboration platform but also scalability and flexibility to support a wide range of collaborative robot studies, such as human trust, emergent behavior, and more. Our preliminary user study results demonstrate the value of the robot assistant in human-robot collaboration, highlighting improvements in overall performance. The observed gap further suggests potential areas for algorithmic advancements. Additionally, the subjective feedback validates the capability of HumanTHOR in facilitating human-related studies, such as assessing human trust. Both quantitative and qualitative findings confirm the effectiveness of our system as a benchmark and experimental platform for human-robot collaboration (HRC) research.

Beyond the conducted experiments, the HumanTHOR system also supports advanced features such as multi-robot settings and more complicated tasks such as collaborative rearrangement. Customized robot algorithms, such as learning-based visual navigation models, are also well supported. In general, our HumanTHOR establishes a foundation and embodied test field for various future work in the human-robot collaboration domain.

References

1. Anderson P, Wu Q, Teney D, Bruce J, Johnson M, Sünderhauf N, Reid I, Gould S, Van Den Hengel A (2018) Vision-and-language navigation: interpreting visually-grounded navigation instructions in real environments. In: Proceedings of the IEEE conference on computer vision and pattern recognition, pp 3674–3683
2. Batra D, Chang AX, Chernova S, Davison AJ, Deng J, Koltun V, Levine S, Malik J, Mordatch I, Mottaghi R, et al (2020) Rearrangement: a challenge for embodied AI. arXiv preprint arXiv:2011.01975
3. Broadbent E, Jayawardena C, Kerse N, Stafford R, MacDonald B (2011) Human-robot interaction research to improve quality of life in elder care: an approach and issues. In: Proceedings of the 12th AAAI conference on human-robot interaction in elder care, pp 13–19
4. Carroll M, Shah R, Ho MK, Griffiths T, Seshia S, Abbeel P, Dragan A (2019) On the utility of learning about humans for human-AI coordination. In: Advances in Neural Information Processing Systems 32 (NeurIPS 2019)
5. Chang A, Dai A, Funkhouser T, Halber M, Niebner M, Savva M, Song, S, Zeng A, Zhang Y (2017) Matterport3d: learning from rGB-D data in indoor environments. In: 2017 International conference on 3D vision (3DV), pp 667–676. IEEE, New York
6. Chao YW, Paxton C, Xiang Y, Yang W, Sundaralingam B, Chen T, Murali A, Cakmak M, Fox D (2022) HandoverSim: a simulation framework and benchmark for human-to-robot object handovers. In: 2022 International conference on robotics and automation (ICRA), pp 6941–6947. IEEE, New York
7. Das A, Datta S, Gkioxari G, Lee S, Parikh D, Batra D (2018) Embodied question answering. In: Proceedings of the IEEE conference on computer vision and pattern recognition, pp 1–10
8. De Pace F, Gorjup G, Bai H, Sanna A, Liarokapis M, Billinghurst M (2021) Leveraging enhanced virtual reality methods and environments for efficient, intuitive, and immersive teleoperation of robots. In: 2021 IEEE international conference on robotics and automation (ICRA), pp 12967–12973. IEEE, New York
9. Deitke M, Han W, Herrasti A, Kembhavi A, Kolve E, Mottaghi R, Salvador J, Schwenk D, VanderBilt E, Wallingford M, et al (2020) Robothor: an open simulation-to-real embodied AI platform. In: Proceedings of the IEEE/CVF conference on computer vision and pattern recognition, pp 3164–3174
10. Devlin J, Chang MW, Lee K, Toutanova K (2018) Bert: Pre-training of deep bidirectional transformers for language understanding. arXiv preprint arXiv:1810.04805
11. Dianatfar M, Latokartano J, Lanz M (2021) Review on existing VR/AR solutions in human–robot collaboration. Procedia CIRP 97:407–411
12. Ehsani K, Han W, Herrasti A, VanderBilt E, Weihs L, Kolve E, Kembhavi A, Mottaghi R (2021) ManipulaTHOR: a framework for visual object manipulation. In: Proceedings of the IEEE/CVF conference on computer vision and pattern recognition, pp 4497–4506
13. Ferrari D, Benzi F, Secchi C (2022) Bidirectional communication control for human-robot collaboration. In: 2022 International conference on robotics and automation (ICRA), pp 7430–7436. IEEE, New York
14. Fu H, Xu W, Ye R, Xue H, Yu Z, Tang T, Li Y, Du W, Zhang J, Lu C (2023) Demonstrating RFUniverse: a multiphysics simulation platform for embodied AI. In: Proceedings of Robotics: Science and Systems. Daegu, Republic of Korea. https://doi.org/10.15607/RSS.2023.XIX.087

15. Fu H, Xu W, Ye R, Xue H, Yu Z, Tang T, Li Y, Du W, Zhang J, Lu C (2023) Demonstrating RFUniverse: A multiphysics simulation platform for embodied AI. In: Robotics: science and systems
16. Gao Q, Thattai G, Gao X, Shakiah S, Pansare S, Sharma V, Sukhatme G, Shi H, Yang B, Zheng D, et al (2023) Alexa Arena: a user-centric interactive platform for embodied AI. arXiv preprint arXiv:2303.01586
17. Gao X, Gao Q, Gong R, Lin K, Thattai G, Sukhatme GS (2022) Dialfred: dialogue-enabled agents for embodied instruction following. IEEE Rob Autom Lett 7(4):10049–10056
18. Gao X, Gong R, Shu T, Xie X, Wang S, Zhu SC (2019) VRKitchen: an interactive 3D virtual environment for task-oriented learning. arXiv preprint arXiv:1903.05757
19. Kaplan KE, Nichols KA, Okamura AM (2016) Toward human-robot collaboration in surgery: performance assessment of human and robotic agents in an inclusion segmentation task. In: 2016 IEEE international conference on robotics and automation (ICRA), pp 723–729. IEEE, New York
20. Kohn S, Blank A, Puljiz D, Zenkel L, Bieber O, Hein B, Franke J (2018) Towards a real-time environment reconstruction for VR-based teleoperation through model segmentation. In: 2018 IEEE/RSJ international conference on intelligent robots and systems (IROS), pp 1–9. IEEE, New York
21. Kolve E, Mottaghi R, Han W, VanderBilt E, Weihs L, Herrasti A, Deitke M, Ehsani K, Gordon D, Zhu Y, et al (2017) AI2-THOR: an interactive 3D environment for visual AI. arXiv preprint arXiv:1712.05474
22. Kumar S, Savur C, Sahin F (2020) Survey of human–robot collaboration in industrial settings: Awareness, intelligence, and compliance. IEEE Trans Syst Man Cybern Syst 51(1):280–297
23. Lee KW, Kim HR, Yoon WC, Yoon YS, Kwon DS (2005) Designing a human-robot interaction framework for home service robot. In: ROMAN 2005. IEEE international workshop on robot and human interactive communication, 2005, pp 286–293. IEEE, New York
24. Li C, Xia F, Martín-Martín R, Lingelbach M, Srivastava S, Shen B, Vainio KE, Gokmen C, Dharan G, Jain T, et al (2022) iGibson 2.0: object-centric simulation for robot learning of everyday household tasks. In: 5th annual conference on robot learning
25. Li X, Fan D, Deng Y, Lei Y, Omalley O (2024) Sensor fusion-based virtual reality for enhanced physical training. Rob Intell Autom 44(1):48–67
26. Li X, Guo D, Liu H, Sun F (2022) Embodied semantic scene graph generation. In: Conference on robot learning, pp 1585–1594. PMLR, New York
27. Liu X, Guo D, Liu H, Sun F (2022) Multi-agent embodied visual semantic navigation with scene prior knowledge. IEEE Rob Autom Lett 7(2):3154–3161
28. Liu X, Li X, Guo D, Tan S, Liu H, Sun F (2022) Embodied multi-agent task planning from ambiguous instruction. In: Proceedings of robotics: science and systems, New York City, NY, USA, pp 1–14
29. Long Y, Wei W, Huang T, Wang Y, Dou Q (2023) Human-in-the-loop embodied intelligence with interactive simulation environment for surgical robot learning. IEEE Rob Autom Lett 8(8):4441–4448
30. Macciò S, Carfì A, Mastrogiovanni F (2022) Mixed reality as communication medium for human-robot collaboration. In: 2022 International conference on robotics and automation (ICRA), pp 2796–2802. IEEE, New York
31. Murphy RR (2004) Human-robot interaction in rescue robotics. IEEE Trans Syst Man Cybern Part C Appl Rev 34(2):138–153
32. Ng E, Liu Z, Kennedy M (2023) It takes two: learning to plan for human-robot cooperative carrying. In: 2023 IEEE international conference on robotics and automation (ICRA), pp 7526–7532. IEEE, New York
33. Padmakumar A, Thomason J, Shrivastava A, Lange P, Narayan-Chen A, Gella S, Piramuthu R, Tur G, Hakkani-Tur D (2022) Teach: Task-driven embodied agents that chat. In: Proceedings of the AAAI conference on artificial intelligence, vol 36, pp 2017–2025

34. Promsutipong K, Luces JVS., Ravankar AA, Tafrishi SA, Hirata Y (2022) Immersive virtual walking system using an avatar robot. In: 2022 international conference on robotics and automation (ICRA), pp 9325–9331. IEEE, New York
35. Qiu S, Liu H, Zhang Z, Zhu Y, Zhu SC (2020) Human-robot interaction in a shared augmented reality workspace. In: 2020 IEEE/RSJ international conference on intelligent robots and systems (IROS), pp 11413–11418. IEEE, New York
36. Rosen E, Whitney D, Fishman M, Ullman D, Tellex S (2020) Mixed reality as a bidirectional communication interface for human-robot interaction. In: 2020 IEEE/RSJ international conference on intelligent robots and systems (IROS), pp 11431–11438. IEEE, New York
37. Sarch G, Fang Z, Harley AW, Schydlo P, Tarr MJ, Gupta S, Fragkiadaki K (2022) TIDEE: tidying up novel rooms using visuo-semantic commonsense priors. In: European conference on computer vision, pp 480–496. Springer, Berlin
38. Savva M, Kadian A, Maksymets O, Zhao Y, Wijmans E, Jain B, Straub, J, Liu J, Koltun V, Malik J, et al (2019) Habitat: a platform for embodied AI research. In: Proceedings of the IEEE/CVF international conference on computer vision, pp 9339–9347
39. Shen B, Xia F, Li C, Martín-Martín R, Fan L, Wang G, Pérez-D'Arpino C, Buch S, Srivastava S, Tchapmi L, et al (2021) iGibson 1.0: a simulation environment for interactive tasks in large realistic scenes. In: 2021 IEEE/RSJ international conference on intelligent robots and systems (IROS), pp 7520–7527. IEEE, New York
40. Shridhar M, Thomason J, Gordon D, Bisk Y, Han W, Mottaghi R, Zettlemoyer L, Fox D (2020) Alfred: a benchmark for interpreting grounded instructions for everyday tasks. In: Proceedings of the IEEE/CVF conference on computer vision and pattern recognition, pp 10740–10749
41. Silva S, Verdezoto N, Paillacho D, Millan-Norman S, Hernández JD (2023) Online social robot navigation in indoor, large and crowded environments. In: 2023 IEEE international conference on robotics and automation (ICRA), pp 9749–9756. IEEE, New York
42. Stotko P, Krumpen S, Schwarz M, Lenz C, Behnke S, Klein R, Weinmann M (2019) A VR system for immersive teleoperation and live exploration with a mobile robot. In: 2019 IEEE/RSJ international conference on intelligent robots and systems (IROS), pp 3630–3637. IEEE, New York
43. Szot A, Clegg A, Undersander E, Wijmans E, Zhao Y, Turner J, Maestre N, Mukadam M, Chaplot DS, Maksymets O, et al (2021) Habitat 2.0: training home assistants to rearrange their habitat. Adv Neural Inf Proces Syst 34:251–266
44. Vasconez JP, Kantor GA, Cheein FAA (2019) Human–robot interaction in agriculture: a survey and current challenges. Biosyst Eng 179:35–48
45. Villani V, Pini F, Leali F, Secchi C (2018) Survey on human–robot collaboration in industrial settings: safety, intuitive interfaces and applications. Mechatronics 55:248–266
46. Wan Y, Mao J, Tenenbaum JB: HandMeThat: human-robot communication in physical and social environments. In: Thirty-sixth conference on neural information processing systems datasets and benchmarks track (2022)
47. Weihs L, Deitke M, Kembhavi A, Mottaghi R (2021) Visual room rearrangement. In: Proceedings of the IEEE/CVF conference on computer vision and pattern recognition, pp 5922–5931
48. Wortsman M, Ehsani K, Rastegari M, Farhadi A, Mottaghi R (2019) Learning to learn how to learn: self-adaptive visual navigation using meta-learning. In: Proceedings of the IEEE/CVF conference on computer vision and pattern recognition, pp 6750–6759
49. Wu Q, Wu CJ, Zhu Y, Joo J (2021) Communicative learning with natural gestures for embodied navigation agents with human-in-the-scene. In: 2021 IEEE/RSJ international conference on intelligent robots and systems (IROS), pp 4095–4102. IEEE, New York
50. Zhang Q, Hu Z, Song Y, Pei J, Liu J (2023) The human gaze helps robots run bravely and efficiently in crowds. In: 2023 IEEE international conference on robotics and automation (ICRA), pp 7540–7546. IEEE, New York

Chapter 8
Application of Embodied Multi-Agent System

Abstract In this chapter, we discuss the real-world applications of the embodied multi-agent system. We employ the proposed framework of the perception-action loop in physical robots in the real-world scenarios. We first introduce the real-world application in the homogeneous embodied multi-agent system for the autonomous task assignment problem and provide the physical validation demonstration for the proposed framework. Then we introduce the real-world application in the heterogeneous embodied multi-agent system for the room tidying-up task. We utilize the quadruped robot with the robotic arm, the small robot only with the navigation ability, and the high robot only with the navigation ability to form the heterogeneous multi-agent system and demonstrate the physical application in the room tidying-up task.

8.1 Application of Homogeneous Embodied Multi-Agent for Autonomous Task Assignment

In this section, we apply the proposed perception-action for homogeneous embodied multi-agent task assignment problem to real-world scenarios.

8.1.1 Deployment Setting

We propose the centralized perception-action framework in task assignment problem in Sect. 3.2.1. As we claim, the proposed framework reduces the gap between the simulation and the real world by using the semantic encoder to represent the visual perception and the subgoal to facilitate navigation. We apply the proposed framework to the real-world environment. We establish a real-world scenario in a practical office building, which is also an unseen environment, but the navigation map is available. This scenario consists of Room1 ($30\,\text{m}^2$), Room2 ($100\,\text{m}^2$), and the corridor (35 m long) connecting them. Following similar guidelines, we design the *Easy* (single room) and *Hard* settings (cross room), which are shown in Fig. 8.1.

H. Liu et al., *Embodied Multi-Agent Systems*, Machine Learning: Foundations, Methodologies, and Applications, https://doi.org/10.1007/978-981-96-5871-8_8

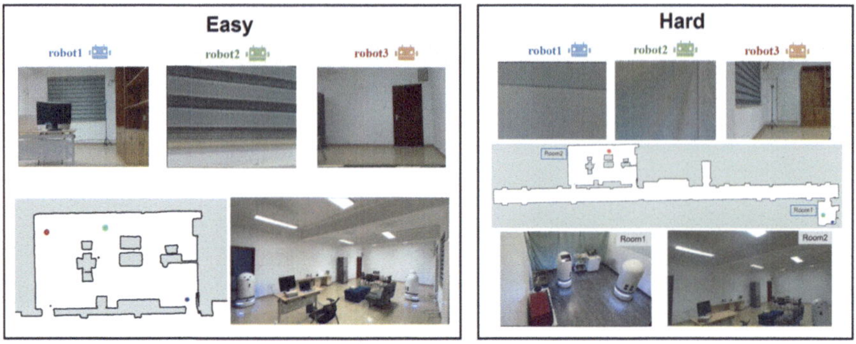

Fig. 8.1 The scenario setup in the real-world application of homogeneous embodied multi-agent autonomous task assignment

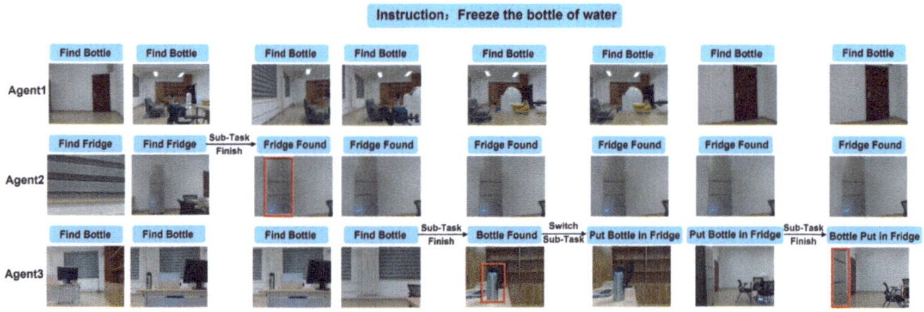

Fig. 8.2 Demonstration of the real-world application of homogeneous multi-agent in task assignment in the *Easy* setting

We deploy three homogeneous customized mobile robots as the embodied agents. Since we use the subgoal navigation policy, the developed method can be directly transferred to the real robots. However, it is noted that our robotic platform does not have the manipulator, and therefore, we assume that the object is virtually manipulated when it is found by the object detector.

8.1.2 Demonstration

We demonstrate the performance of the real-world application of the homogeneous multi-agent task assignment framework in the *Easy* setting in a single room in Fig. 8.2. Agents perform the task *"Freeze the bottle of water"*. Agent 3 finds the bottle of water, then finds the fridge with the help of communication information from Agent 1 and Agent 2, and finally completes the task successfully.

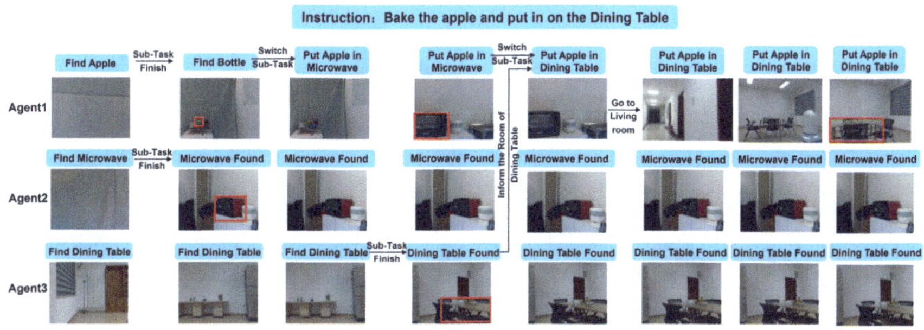

Fig. 8.3 Demonstration of the real-world application of homogeneous multi-agent in task assignment in the *Hard* setting

In Fig. 8.3, we demonstrate the process of the homogeneous multi-agent task assignment in the *Hard* setting. Agents perform the task *"Bake the apple and put it on the dining table."* Agent 1 and Agent 2 collaborate to find the apple and the microwave to bake the apple in Room1. Agent 3 finds the dining table in Room2. Then Agent 1 decides to move to Room2 from Room1 to put the baked apple on the dining table in Room2. After Agent 1 enters Room1, it finds the dining table based on the communication messages from Agent3 and finally completes this task successfully. From the experiments in the real-world scenes, it is found that the encoded semantic map and the subgoal prediction in the proposed framework can help shrink the gap in visual and action domains between simulation and the real world.

8.2 Application of Heterogeneous Embodied Multi-Agent for Room Tidying-up

In this section, we apply the proposed perception-action for heterogeneous embodied multi-agent room tidying-up task to the real-world scenarios.

8.2.1 Deployment Setting

We propose the LLM-based ad hoc framework in heterogeneous multi-agent room tidying-up task in Sect. 6.3. Due to the remarkable generalization ability of the LLM with the proposed IRoT method, the proposed ad hoc framework can be applied to the real-world scenario. In the real-world application, we utilize three heterogeneous robots for the experimental procedure in the tidying-up task, as illustrated in Fig. 8.4a. It is crucial to highlight that these three agents demonstrate considerable differences in their visual range and manipulation capacities. The agent on the far

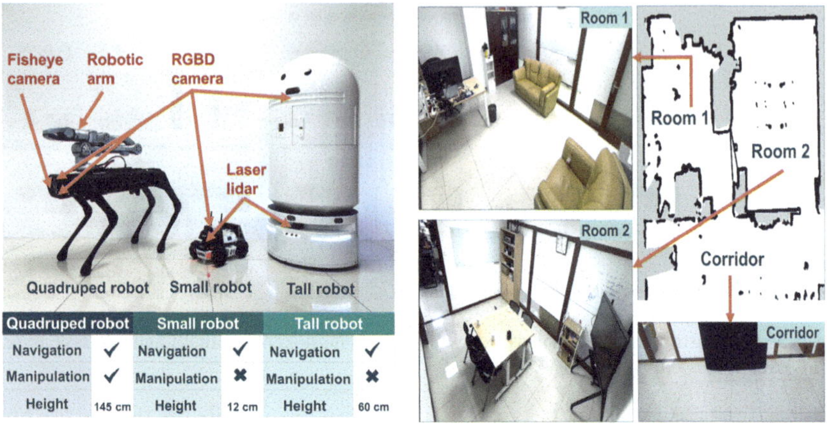

(a) (b)

Fig. 8.4 The robots and environment settings in the real-world application of heterogeneous embodied multi-agent room tidying-up task. (**a**) Robots. (**b**) Environments

left is a quadruped robot with a manipulator, followed by a small four-wheeled robot in the center, and the last agent is a tall robot engineered for indoor navigation. The experimental environment is configured as two rooms connected by a corridor, as shown in Fig. 8.4b. In the real-world experiments, each robot has a computer to process the information from the RGB-D camera and the Laser Radar, record its movement path, and control the robot to execute the corresponding actions. There also exists another computer to cache the communication messages and act as the intermediate node to exchange communication information among robots. Each robot records its trajectory in real time. The average latency for receiving responses from LLMs is approximately two seconds when planning the sub-task and the subskill. Unlike the simulation experiment, where agents synchronously execute actions at each timestep, in real-world experiments, robots execute actions asynchronously. It allows robots to continue executing actions, while others are waiting for the LLM response.

8.2.2 Demonstration

We establish an original team consisting of the quadruped robot and the tall robot, introducing the small robot into the environment after the task starts for a period of time. The original team utilizes the *LLM-based Policy* as the collaboration strategy, and the ad hoc robot utilizes the *IRoT-LLM Policy*. As shown in Fig. 8.5, during this task, the quadruped robot identifies and picks up an empty bottle from the sofa and then puts it in the trash can. When no ad hoc robot joins the original team, the quadruped robot needs to search and locate the misplaced full bottle itself

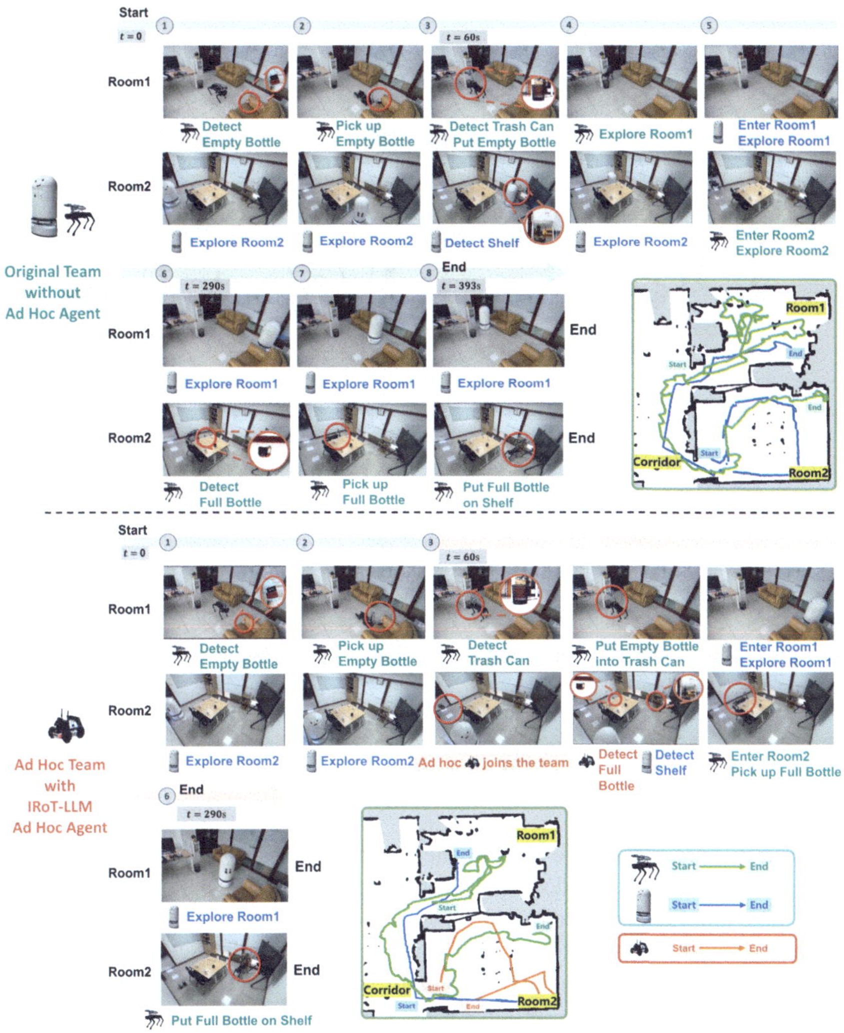

Fig. 8.5 Demonstration of the ad hoc robot collaborating with the LLM-based original team in the real-world experiment

after it completely explores *Room1*, because the tall robot cannot see the full bottle on the floor. However, when the small robot joins the original team as an ad hoc robot after 60 seconds, it can detect the full bottle sooner and send this message to the quadruped robot, which helps the quadruped robot find the full bottle and put it on the shelf faster. The whole task finishes once the quadruped robot puts the full bottle back to the shelf. The execution process of this task demonstrates that with the assistance of the ad hoc robot, the efficiency of the task is obviously

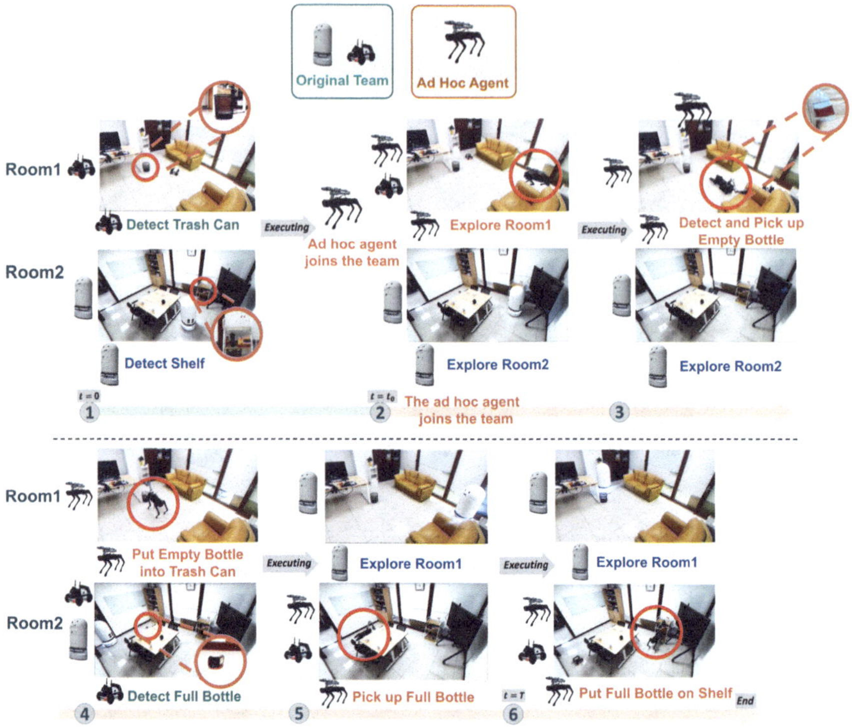

Fig. 8.6 Demonstration of the extended setting in the real-world application

increased compared with the performance of the original team. The paths of the original teammates are also decreased compared with the original paths.

We also consider another extended setting in which the original team cannot complete the task in the real-world scenario in Fig. 8.6. In this setting, the original team consists of the tall robot and the small robot, and the quadruped robot serves as the ad hoc robot. After the original team starts the task for a period of time, the quadruped robot joins the team to collaborate with other robots. Since all robots in the original team do not have the manipulation ability, they cannot complete this task. After the quadruped robot joins the team, the new team can complete this task with information from the original team. The quadruped robot picks up the empty bottle and puts it to the trash can, and then it enters the *Room2* to pick up the full bottle with the communication information from the small robot and puts it on the shelf. This sample demonstrates that the proposed ad hoc framework can not only solve the ad hoc teamwork task in the case that the original team can complete the task, but also can handle the case that the original team does not have the ability to complete the task.

Part IV
Conclusions

This part of this book comprises one chapter, which summarizes the main work of this book and presents some important future directions.

Chapter 9
Conclusions and Future Directions

9.1 Summary

In this book, we systematically discussed the background, theory and methods, and selected applications of embodied multi-agent system. Specially, in Part I, we first reviewed the history of embodied intelligence and presented a general architecture about it. Then, we provided a general introduction to the more specific topic of embodied multi-agent system. Followed by them, we presented a series of methods to tackle the perception-action loop, embodied cooperation, and competitive learning. Finally, we presented some advanced applications including large language models, simulation platforms, and physical robotic validation. Since our work focuses on the embodied multi-agent system, we are especially interested in the synergy of perception, action, and learning, which has also been emphasized in the first chapter. Therefore, the presented content significantly differs from the existing work on coordinated control, collaborative perception, and swarm intelligence, which are also important topics in multi-agent. We hope such a viewpoint could provide a new perspective for the researchers in this field. Nevertheless, we admit that there still exists a great gap between perception, action, and learning, and even morphology. And more investigation should be performed to bridge the gap.

9.2 Future Directions

9.2.1 Active Collaboration

Though many works have been developed to perform the embodied collaboration, most of them assume a prior configuration is given and kept fixed. This assumption is too strict for practical scenarios and cannot deal with many emergent situations. In this book, we systematically develop the architectures of embodied collaboration for

H. Liu et al., *Embodied Multi-Agent Systems*, Machine Learning: Foundations, Methodologies, and Applications, https://doi.org/10.1007/978-981-96-5871-8_9

homogeneous and heterogeneous embodied agents and make a preliminary attempt to solve the ad hoc collaboration, which is a very challenging problem. In addition, the ad hoc collaboration has been proposed for more than 10 years and the research progress is slow. The main reason is the complicated intrinsic nature of the problem setting and the lack of a suitable modeling tool. Very recently, the emergence of LLM has opened the doors for many complicated problems, and we find LLM also presents great potential for embodied multi-agent system. Nevertheless, we find the current work still limited the scenarios in which the configurations are passively changed. That is to say, the agents always try to be adaptive to the configuration changing but cannot actively ask for help from other agents. This is a serious drawback of the current work. In fact, asking for help is a very important intelligent behavior, and it also provides a solution for active collaboration.

9.2.2 Interactive Learning

In multi-robot systems, each robot is typically equipped with diverse sensors (e.g., cameras, radar) to acquire multimodal information. Through information sharing and collaborative interactions among robots, robots not only augment their own sensing and decision-making capabilities but also harness the collective intelligence of the system, enabling them to tackle complex environmental challenges and meet diverse operational requirements.

Within the framework of multi-robot collaborative learning, robots share sensory information (e.g., visual images, infrared signals, point clouds) to construct a comprehensive understanding of their environment. For example, in the rescue mission, ground robots are equipped with camera and radar sensors to capture environmental information, while aerial robots are equipped with infrared sensors to detect thermal signatures. By integrating these multimodal data, the system is able to accurately identify survivors, assess hazardous areas, and detect obstacles within the disaster zone, thereby facilitating the efficiency of collaboration during rescue scenes. The mode of multimodal collaborative learning not only improves the perceptual capabilities of the system but also enhances its robustness in performing complex tasks.

Cross-modal learning is an extension of multimodal learning and holds the potential to significantly enhance the performance and adaptability of multi-robot systems, particularly in complex environments. Within the cross-modal learning framework, the system not only processes multimodal data but also understands the interdependencies between different modalities. Moreover, it has the capability of inferring or supplementing data from other modalities. For example, poor lighting conditions may result in blurred visual images, making it difficult to identify the target object accurately. However, point cloud data from a robot located at a different viewpoint can provide precise information about the location and movement of the target object. By complementing different modal data from multiple robots, the system can infer the category of the target object. Cross-modal learning builds deep connections between different modalities, facilitating data complementarity and significantly improving the efficiency and accuracy of the task.

9.2.3 Embodied Human-Agent Collaboration

In many scenarios, the reliability of robots is not very good, and they often exhibit strange unexpected behavior. All of those significantly prevent robots from successful application. Currently, the most effective method to solve this problem is to introduce the human in the loop. This is not a new concept and has been studied for many years. Generally speaking, we can divide the research into three categories. The first one is robot-centric. In this case, the human serves the robot, and the robot performs most of the tasks. The representative example is the robot that is teleoperated by the human. The second one is human-centric. In this case, the human does most of the tasks and the robot provides necessary helps. A representative example is the exoskeleton, in which the robot provides extra strength, enabling the human to complete more difficult tasks. For both cases, we can see the clear boundary between the human and the robot, because one serves as a master and the other a slave. In fact, with the development of robots, the boundary between humans and robots is not so clear, and in many cases, a better manner is to establish a team of humans and robots, in which they collaborate equally. The goal of inventing robots is not to replace humans, but to become partners of humans. However, as we have mentioned, since the robots are not so reliable, and human also often makes mistake. Then, can the human trust the robot, or can the robot trust the human? This is also a very challenging problem. To systematically address the human-agent collaboration, we should seriously consider the effective commutation problem in human-automation collaboration, the trust modeling problem in human-robot collaboration, and the Theory-of-Mind in Human-AI collaboration.

The manufacturer's authorised representative in the EU is Springer Nature Customer Service Centre GmbH, Europaplatz 3, 69115 Heidelberg, Germany. If you have any concerns regarding our products, please contact ProductSafety@springernature.com

Printed and bound by CPI Group (UK) Ltd, Croydon, CR0 4YY
01/10/2026
02225930-0002